T0216949

SpringerBriefs in Computer Science

SpringerBriefs present concise summaries of cutting-edge research and practical applications across a wide spectrum of fields. Featuring compact volumes of 50 to 125 pages, the series covers a range of content from professional to academic.

Typical topics might include:

- A timely report of state-of-the art analytical techniques
- A bridge between new research results, as published in journal articles, and a contextual literature review
- A snapshot of a hot or emerging topic
- An in-depth case study or clinical example
- A presentation of core concepts that students must understand in order to make independent contributions

Briefs allow authors to present their ideas and readers to absorb them with minimal time investment. Briefs will be published as part of Springer's eBook collection, with millions of users worldwide. In addition, Briefs will be available for individual print and electronic purchase. Briefs are characterized by fast, global electronic dissemination, standard publishing contracts, easy-to-use manuscript preparation and formatting guidelines, and expedited production schedules. We aim for publication 8–12 weeks after acceptance. Both solicited and unsolicited manuscripts are considered for publication in this series.

**Indexing: This series is indexed in Scopus, Ei-Compendex, and zbMATH **

David Ryckelynck · Fabien Casenave ·
Nissrine Akkari

Manifold Learning

Model Reduction in Engineering

 Springer

David Ryckelynck
Centre des Matériaux
Mines Paris—PSL
Évry, France

Nissrine Akkari
Department of Digital Sciences
and Technologies
Safran Tech
Châteaufort, France

Fabien Casenave
Department of Digital Sciences
and Technologies
Safran Tech
Châteaufort, France

ISSN 2191-5768   ISSN 2191-5776 (electronic)
SpringerBriefs in Computer Science
ISBN 978-3-031-52766-1   ISBN 978-3-031-52764-7 (eBook)
https://doi.org/10.1007/978-3-031-52764-7

This research was partially funded by the French Fonds Unique Interministériel (MOR_DICUS).

© The Editor(s) (if applicable) and The Author(s) 2024. This book is an Open Access publication.

Open Access This book is licensed under the terms of the Creative Commons Attribution 4.0 International License (http://creativecommons.org/licenses/by/4.0/), which permits use, sharing, adaptation, distribution and reproduction in any medium or format, as long as you give appropriate credit to the original author(s) and the source, provide a link to the Creative Commons license and indicate if changes were made.
The images or other third party material in this book are included in the book's Creative Commons license, unless indicated otherwise in a credit line to the material. If material is not included in the book's Creative Commons license and your intended use is not permitted by statutory regulation or exceeds the permitted use, you will need to obtain permission directly from the copyright holder.
The use of general descriptive names, registered names, trademarks, service marks, etc. in this publication does not imply, even in the absence of a specific statement, that such names are exempt from the relevant protective laws and regulations and therefore free for general use.
The publisher, the authors, and the editors are safe to assume that the advice and information in this book are believed to be true and accurate at the date of publication. Neither the publisher nor the authors or the editors give a warranty, expressed or implied, with respect to the material contained herein or for any errors or omissions that may have been made. The publisher remains neutral with regard to jurisdictional claims in published maps and institutional affiliations.

This Springer imprint is published by the registered company Springer Nature Switzerland AG
The registered company address is: Gewerbestrasse 11, 6330 Cham, Switzerland

Paper in this product is recyclable.

Foreword

Safran Tech is the corporate research center of Safran Group, a French multinational company that designs, develops and manufactures aircraft and rocket engines, as well as various aerospace and defense-related equipment. This book is the result of the collaboration of two of my research engineers at SafranTech, Fabien Casenave and Nissrine Akkari, with David Ryckelynck, professor of Applied Mathematics at MINES Paris PSL, a French engineering school. This collaboration dates back to the creation of Safran Tech in 2015 and has involved Ph.D. students, namely Thomas Daniel and Hamza Boukraichi. Thomas Daniel won the 2022 AMIES Ph.D. award (AMIES: Agency for Mathematics in collaboration with companies and the society) for his work on dictionary-based ROM-nets, which are introduced in this book.

Nowadays, the engineering and design processes involved when creating a new mechanical part, component or product in industrial companies like Safran rely on some level of numerical simulation. In some cases, complex and high-dimensional models are required to be computed, and often in a so-called "many queries" context: when searching for an optimal design, quantifying uncertainties or exploiting a "real-time" simulator, the physics problem is solved a large number of time at different configurations. Hence, we need fast surrogates, with a control of the approximation error. Among the available families of methods, this book deals with *projection-based reduced-order models*, which consists in solving the same physics equations as the reference high-dimensional models, but on a reduced dimensional vector space. Resorting to the known physics equations provides explainability, error estimates in some cases, and good performance with sparse learning databases. Classical and modern (e.g., deep learning) machine learning technologies are used to assist *projection-based reduced-order models*, either in annex tasks like clustering or classification or by providing non-linear dimensionality reduction procedures. Challenges of these methods include geometrical variabilities, predictive uncertainty quantification, and efficiency (i.e., speed-up factor, also taking the learning stage into account).

This book is addressed to graduate students and researchers who look for an introduction to *projection-based reduced-order models*, an understanding of the main features and challenges of these methods, as well as its state of the art application to real-life engineering design problems and its extensions.

Paris, France   Christian Rey
September 2023

Acknowledgements We acknowledge the contribution from the many co-authors of our publications, in particular Thomas Daniel and Christian Rey for their help in Chap. 5.

Christian Rey is head of the *Mathematics and Algorithms for Design by Simulation* research unit at SafranTech and professor of Computational Mechanics. Christian is a long-time collaborator of the authors of this book.

Contents

Acronyms

AI	Artificial Intelligence
BPIM	Best Point Interpolation Method
CAD	Computer-Aided Design
CNN	Convolutional Neural Networks
DAE	Deep AutoEncoder
(D)EIM	(Discrete) Empirical Interpolation Method
DoE	Design of Experiments
ECM	Empirical Cubature Method
ECSW	Energy-Conserving Sampling and Weighting
FE	Finite Elements
FETI	Finite Element Tearing and Interconnect
FFT	Fast Fourier Transform
GAN	Generative Adversarial Networks
GNAT	Gauss-Newton with Approximated Tensors
HFM	High-Fidelity Model
HP	High-Pressure (turbine blade)
HPC	High-Performance Computing
LES	Large Eddy Simulation
LHS	Latin Hypercube Sampling
LPEQ	Linear Program Empirical Quadrature Procedure
MBE	Model-Based Engineering
MDS	Multidimensional Scaling
MMGP	Mesh Morphing Gaussian Process
MPE	Missing Point Estimation
mRMR	Minimum Redundancy Maximum Relevance
NDT	Non Destructive Testing
PAM	Partitioning Around Medoids
PDE	Partial Differential Equations
PINN	Physics-Informed Neural Network
PLM	Product Lifecycle Management
POD	Proper Orthogonal Decomposition

PVC	Precessing Vortex Core
RID	Reduced Integration Domain
ROB	Reduced-Order Basis
ROM	Reduced-Order Model
SMACOF	Scaling by MAjorizing a COmplicated Function
SVD	Singular Value Decomposition
TBC	Thermal Barrier Coating
TKE	Turbulent Kinetic Energy
TLV	Tight Local Volumes
UQ	Uncertainty Quantification
VAE	Variational AutoEncoder

Chapter 1
Structured Data and Knowledge in Model-Based Engineering

1.1 Nomenclature

In the book, the following notation is used: bold face symbols denote vectors (lowercase letters) $\mathbf{a}$, $\mathbf{b}$ ... or matrices (uppercase letters) $\mathbf{A}$, $\mathbf{B}$... Data are denoted by $\mathbf{X}$ for input data and labels or output data are denoted by $\mathbf{Y}$. The spatial gradient operator is denoted by ∇, and the divergence is expressed by $\nabla \cdot \bullet$. The dependence on spatial coordinates $\boldsymbol{\xi}$ is most of the time omitted for simplicity of notation. Modeling parameters are denoted by $\boldsymbol{\mu}$ or $\boldsymbol{\mu}_\square$ when they have a tensor shape (multiple indices). Weights in neural networks are denoted by $\mathbf{W}$. In mechanical models, the Cauchy stress is denoted by $\boldsymbol{\sigma}_c$. Solutions of physics-based models are denoted by $\widetilde{u}$ for partial differential equations (PDEs) or ordinary differential equations (ODEs). High-fidelity solutions in finite dimensional space are denoted by u. Solutions in reduced dimensional spaces are denoted by $\widehat{u}$. PDEs are set up over a domain denoted by Ω. Solution spaces are the following:

- $\mathcal{V}$ for PDEs,
- $\mathcal{V}_h$ for high fidelity solution space of dimension N,
- $\mathcal{V}_n$ for reduced vector space of dimension $n < N$.

The finite element shape functions are denoted by φ. The coordinates of a finite element solution are denoted by the vector $\mathbf{u}$. The reduced vector space $\mathcal{V}_n$ is spanned by empirical modes denoted by $(\psi_k)_{k=1,\dots,n}$. The matrix form of a reduced basis is denoted by $\mathbf{V} \in \mathbb{R}^{N \times n}$. The vector of reduced coordinates is denoted by $\boldsymbol{\gamma} \in \mathbb{R}^n$; $\gamma_k, k \in [1, \dots, n]$ is the kth component of $\boldsymbol{\gamma}$. When needed, variables depending on parameter $\boldsymbol{\mu}$ are denoted using an exponent $\bullet^\mu$ or an index $\bullet_\mu$.

© The Author(s) 2024

D. Ryckelynck et al., *Manifold Learning*, SpringerBriefs in Computer Science, https://doi.org/10.1007/978-3-031-52764-7_1

1.2 Model-Based Engineering

Today, a large set of engineering tasks is supported by mathematical models related to various scientific disciplines. This set of tasks is called **Model-Based Engineering** (MBE). In this book, we restrict our attention to geometrical or morphological models, thermal models, mechanical models and statistical models, at various scales. The related mathematical equations can be algebraic equations, ordinary differential equations, partial differential equations (PDEs), or every possible combination of these types of equations in a coupled system of equations.

Various numerical methods have been developed in applied mathematics, to find approximate solutions of these different types of equations. Therefore, mathematical properties are available related to the convergence of the numerical solutions to the exact solutions having no numerical approximation. The engineers have a strong confidence in the numerical predictions obtained by these models, provided that they are well used in their domain of validity. Hence, in practice, numerical models and numerical predictions are ubiquitous in MBE.

In industrial sectors, such as automotive industry, aeronautical industry, energy, shipbuilding, electronics manufacturing, etc., **computer platforms for MBE** have been developed for assemblies of components, to ease the communication of data and models, between engineering tasks. Quality and efficiency in engineering tasks depend directly on the **continuity of data** on computer platforms. As explained in the ATLAS program of AFNet,[1] such continuity of data requires the development of standards for data exchanges between stakeholders in a project, a production line, a department, etc. Assemblies of components that are equipped with numerical models are termed complex systems in this book.

A **complex system** is an assembly of components that have a numerical representation related to coupled physics-based models, for its design phase, its manufacturing phase or its exploitation phase.

In computer platforms for MBE, complex systems have an idealized digital representation, termed **digital mock-up**, where model couplings can be implemented in a numerical model dedicated to an engineering task. For instance, in continuum mechanics, the mechanical model of a component is weakly coupled to the geometrical model of this component. Many multi-physics couplings in coupled problems, can be implemented on computer platforms for MBE. Complex numerical simulations in MBE are supported by computers that range form laptops to computer clusters for high-performance computing (HPC). In most cases, geometrical models are created by using computer-aided design (CAD) in order to obtain a common geometric model for both design tasks and manufacturing tasks, of each part and assemblies in a complex system.

Data flux in computer platforms for MBE are related to design tasks, manufacturing tasks, non destructive testing (NDT) tasks, exploitation or monitoring tasks, inspection tasks, repairing tasks, recycling tasks. Figure 1.1 is a summary of these data fluxes.

---

[1] https://www.afnet.fr/en/.

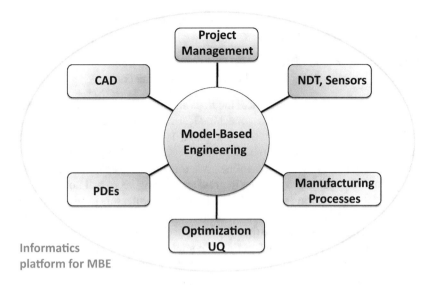

Fig. 1.1 Schematic view of data continuity in an computer platform for Model-Based Engineering (MBE), enabling engineering tasks related to CAD, PDE solution

The value of data, beyond the tasks at hand, should not be ignored by engineers. In particular, the value of data must be considered in light of advances in artificial intelligence (AI) and machine learning methods.

1.3 Motivation

Our main paradigm is: the more we know about underlying mathematical properties or physical properties of data, the less we need data for machine learning tasks. In this book, we aim to learn manifold from data and conventional knowledge in engineering. This paradigm has been recently employed in Physics Informed Neural Networks [6, 7] and knowledge transfer [4]. Similarly to transfer learning, we are "motivated by the fact that people can intelligently apply knowledge learned previously to solve new problems faster or with better solutions" [5].

Much data in computer platforms for MBE are structured data, in the sense that their format is predefined and documented (this notion is different from structured and unstructured meshes, referring respectively to constant rectilinear and heterogeneous meshes). These structured data are supported by a graph or they have a tensor format, like digital images. Since deep learning has made huge progresses in computer vision, image transformation and for graph neural networks, there is an opportunity to develop deep learning in MBE. The development of sensors and image acquisitions in non destructive testing of complex systems makes this opportunity more and more relevant. The connection of computer platforms to HPC facilities leads to the two following issues:

- Where are the data in MBE?
- What can be learned that is useful or valuable to engineers?

In this book, we propose numerical methods and algorithms to assist engineers in various conventional tasks by using machine learning. This gives value to data that flow in computer platforms for MBE. Our goal is to augment engineers capabilities by using artificial intelligence, without ignoring the knowledge already available for engineering in all scientific disciplines. We think that the collaboration between engineers and data scientists is going to grow up, but engineers will not be replaced by them. At the engineering school MINES Paris PSL, professors David Ryckelynck and Elie Hachem created in 2017 master courses to ease such collaborations between engineers and data scientists. The title of this set of courses is Ingénierie Digitale des Systèmes complexes (Digital engineering of complex systems). The creation of such a course would not have been possible without the help of SAFRAN, both for the illustrations of classical AI and for the development of an engineering augmented by a hybrid AI, which preserves scientific knowledge.

Acceleration of engineering tasks is the main objective here, but we are also able to develop engineering tasks that were unaffordable without machine learning. This is especially true when considering uncertainty quantification (UQ) or image-based digital twining of complex systems.

Following the definition in [2], a digital twin involves similar numerical models to mock-up but it is related to an existing instance of a complex system or an instance of one of its component. A digital twin also include the related observational data. Hence, in essence, a digital twin has **multimodal data**, such as observational data and simulated data. The acceleration of image-based modeling for digital twining of manufactured parts is going to foster the development of functional assessment, in which defects harmfulness are evaluated with respect to physics-based of mechanical-based specifications.

The knowledge accumulated via digital twining or via sensors only, enable the development of uncertainty quantification. Sensors and numerical mock-up being more and more spatially resolved and accurate, the acceleration of numerical predictions is becoming crucial for practical application in engineering. In Sect. 5, mechanical predictions are accelerated by using machine learning in order to account for stochastic variation of temperature fields when computing the lifetime of turbine blades.

In this book we develop the following machine learning tasks, via manifold learning:

- classification,
- anomaly detection,
- regression,
- dimensionality reduction.

Data visualization is out of the scope of this book, although dedicated manifold learning techniques have been proposed in the literature. In this book, manifold learning is thought of as an attempt to generalize linear frameworks for numerical methods in engineering.

When accumulating data in computer platform for MBE, data do not fill the entire ambient space used for the encoding of these data. There exists an empirical space, or a **latent space**, occupied by data whose dimension is usually much smaller than the ambient space. Dimensionality reduction and manifold learning aim to find a numerical approximation of this latent space. As models are ubiquitous, we have to consider large sets of synthetic data, or more precisely, sets of simulated data for engineering tasks. These simulated data are solutions of numerical equations. They belong to a **solution space**, which is the ambient space for simulated data. A solution space is directly related to the general numerical scheme used to compute instances of numerical predictions. Given an engineering task, an ambient space is a common space for all data instances. It does not depend on any variable parameter. The dimension of the ambient space is set up so that each instance of data is a point inside the ambient space.

Two kinds of latent spaces can be defined when considering simulated data:

- latent space for surrogate modeling,
- latent space for projection-based model order reduction.

Projection-based model order reduction is detailed in Chap. 2. Our recommendation for selecting a surrogate model rather than a projection-based reduced order model is the following. If you consider a mature engineering task so that:

- you can decide a priori quantities of interest,
- all influencing parameters have known range of variation, and if possible a known probability density functions,
- you are able to sample the parameter space without suffering from the curse of dimensionality and build a data-base,
- you need real-time predictions,

then you should use a surrogate modeling rather a projection-based reduced order modeling. But, when modeling a complex system by using PDEs, the selection of simulation outcomes is rarely trivial. Generally speaking, projection-based reduced order models are slower than surrogate models, but one can adapt the outcomes of reduced predictions and their validity domain is not restricted to the parameter domain used to learn the reduced order model, both in term of extent and in term of dimension.

In practice, data accumulation has to be supplemented by data compression schemes but also data pruning scheme. As explained in [3], manifold learning for projection-based model order reduction is a convenient way for data pruning that preserves modeling capabilities. Pruning algorithms facilitate data augmentation for high-dimensional data, by limiting the memory requirements for augmented data [1].

So, data in MBE are simulated data and observational data, where simulated data are synthetic data forecast by physics-based models. Unfortunately, observational data on complex systems are extremely rare in engineering. For instance, aircraft landing gear tests, automotive crash tests, etc. are mostly done for conformity assessment processes or for certification processes. For example, in the aeronautics industry, aircraft manufacturers must demonstrate compliance of new products with regulatory requirements. This compliance demonstration is done by analysis during ground testing (such as tests on the structure to withstand bird strikes, fatigue tests and tests in simulators) but also by means of tests during flight. Usually, manufacturers have many more instances of observational data, at the components scale, during manufacturing processes, maintenance and exploitation tasks. In engineering, the product lifecycle management (PLM) refers to the management of data and processes across the entire lifecycle of a component or a product. PLM requires a high level of data continuity across the supply chain, by developing data standards and dedicated computer platforms. Such computer platforms open the route for more machine learning in engineering, especially in MBE where engineers are facing very complex structured and unstructured data.

In this book we restrict our attention to spatially structured data. The detailed description of the data we worked with are available in Chap. 6.

References

1. A. Aublet, F. N'Guyen, H. Proudhon, D. Ryckelynck, Multimodal data augmentation for digital twining assisted by artificial intelligence in mechanics of materials. Front. Mater. **9** (2022)
2. E.H. Glaessgen, D. Stargel, *The digital twin paradigm for future Nasa and US Air Force vehicles, in The 53rd Structural Dynamics, and Materials Conference: Special Session on Digital Twin* (HI, USA, Honolulu, 2012), pp.1–14
3. W. Hilth, D. Ryckelynck, C. Menet, Data pruning of tomographic data for the calibration of strain localization models. Math. Comput. Appl. **24**(1) (2019)
4. A.T.W. Min, R. Sagarna, A. Gupta, Y.-S. Ong, C.K. Goh, Knowledge transfer through machine learning in aircraft design. IEEE Comput. Intell. Mag. **12**(4), 48–60 (2017)
5. S.J. Pan, Q. Yang, A survey on transfer learning. IEEE Trans. Knowl. Data Eng. **22**(10), 1345–1359 (2010)
6. M. Raissi, P. Perdikaris, G.E. Karniadakis, Physics informed deep learning (part I): Data-driven solutions of nonlinear partial differential equations. ArXiv preprint (2017) arXiv:1711.10561
7. M. Raissi, P. Perdikaris, G.E. Karniadakis, Physics informed deep learning (part II): Data-driven discovery of nonlinear partial differential equations. ArXiv preprint (2017) arXiv:1711.10566

Open Access This chapter is licensed under the terms of the Creative Commons Attribution 4.0 International License (http://creativecommons.org/licenses/by/4.0/), which permits use, sharing, adaptation, distribution and reproduction in any medium or format, as long as you give appropriate credit to the original author(s) and the source, provide a link to the Creative Commons license and indicate if changes were made.

The images or other third party material in this chapter are included in the chapter's Creative Commons license, unless indicated otherwise in a credit line to the material. If material is not included in the chapter's Creative Commons license and your intended use is not permitted by statutory regulation or exceeds the permitted use, you will need to obtain permission directly from the copyright holder.

Chapter 2
Learning Projection-Based Reduced-Order Models

2.1 Motivation and Basic Assumptions

In this chapter, we introduce the solution space for high-fidelity models based on partial differential equations and the finite element model. The manifold learning approach to model order reduction requires simulated data. Hence, learning projection-based reduced order models (ROM) has two steps: (i) an offline step for the computation of simulated data and for consecutive machine learning tasks, (ii) an online step where the reduced order model is used as a surrogate for the high fidelity model. The offline step generates a train set and a validation set of simulated data. The accuracy and the generalisation of the reduced order model is evaluated in the online step by using a test set of data forecast by the high-fidelity model. The test set aims also to check the computational speedups of the reduced-order model compare to the high-fidelity model.

Learning projection-based reduced order model makes sense only if there is a significant computational speedup at the price of an acceptable loss of accuracy in predictions. The longer the computational time of the high-fidelity model, the smaller the acceptable speed up, if we save hours or days of numerical simulations. Regarding the acceptable accuracy of reduced predictions, engineers working in industry and scientists working in laboratories do not have the same expectations. In our own experience, learning projection-based reduced order model has the capability to adapt to engineering tasks, high-fidelity models elaborated in laboratories, in terms of accuracy and computational time. This approach contributes to data continuity between physics-based knowledge developed in laboratories and practical applications for engineering tasks.

From the mathematical point of view, Céa's lemma gives an overview of manifold learning for model order reduction applied to elliptic equations. Let $\mathcal{V}$ be a real Hilbert space, such that the weak form of the elliptic equation reads: find $\widetilde{u} \in \mathcal{V}$, such that:

$$a(\widetilde{u}, \widetilde{v}) = L(\widetilde{v}), \quad \forall \widetilde{v} \in \mathcal{V}, \tag{2.1}$$

© The Author(s) 2024
D. Ryckelynck et al., *Manifold Learning*, SpringerBriefs in Computer Science,
https://doi.org/10.1007/978-3-031-52764-7_2

where $a(\cdot, \cdot)$ is a bilinear form, with coercivity constant $\beta > 0$ and continuity continuity constant $C^a > 0$. $L(\cdot)$ is a linear form. Section 2.2 gives more details on weak forms. Let us $\mathcal{V}_h$ a finite dimensional subspace of $\mathcal{V}$. Here, $\mathcal{V}_h$ is an approximate solution space for the elliptic equation. The approximate solution of the elliptic equation is denoted by $u \in \mathcal{V}_h \subset \mathcal{V}$. The Galerkin projection of the elliptic equation onto the solution space reads: find $u \in \mathcal{V}_h \subset \mathcal{V}$ such that:

$$a(u, v) = L(v), \quad \forall \, v \in \mathcal{V}_h, \tag{2.2}$$

where $\mathcal{V}_h$ has been substituted for $\mathcal{V}$. Céa's lemma states that:

$$\|\widetilde{u} - u\| \leq \frac{C^a}{\beta} \, \|\widetilde{u} - v\|, \quad \forall \, v \in \mathcal{V}_h, \tag{2.3}$$

where $\| \cdot \|$ is a norm in $\mathcal{V}$. The related scalar product is denoted by $\langle \cdot \rangle$. In Finite Element models, $\mathcal{V}_h$ is the span of finite element shape functions. But Céa's lemma holds for all finite-dimensional subspace of $\mathcal{V}$. The closer $\widetilde{u}$ to the solution space $\mathcal{V}_h$, the smaller the right-hand term of Céa's lemma (2.3), and therefore one can expect a better prediction u in Eq. (2.2), although we do not know $\widetilde{u}$.

In few words, a reduced-order model is obtained by introducing a smaller solution space $\mathcal{V}_n \subset \mathcal{V}_h$ of smaller dimension $n < N$. A projection-based reduced order model can be achieved by using the Galerkin projection (2.2), where $\mathcal{V}_n$ is substituted for $\mathcal{V}_h$. The conclusion of Céa's lemma holds again. The closer $\widetilde{u}$ to the solution space $\mathcal{V}_n$, the better the prediction $\widehat{u} \in \mathcal{V}_n$. Manifold learning comes into play when we are given a set of predictions $(u^{(i)})_{i=1,\ldots,m}$ in a common ambient space $\mathcal{V}_h$, related to a given finite element mesh. The basic assumptions in manifold learning for model order reduction are:

- a latent space of reduced dimension is hidden in the data $(u^{(i)})_{i=1,\ldots,m}$, its dimension is denoted by n,
- a machine learning algorithm is available to learn this latent space by using a train set of simulated data extracted from $(u^{(i)})_{i=1,\ldots,m}$,
- the distance between $\widetilde{u}$ and this latent space is small enough, although we do not know $\widetilde{u}$,
- a numerical scheme enables the projection of the elliptic equation onto the latent space, in order to set up the reduced order model,
- the computational complexity of the solution of the reduced order model is smaller that the computational complexity of the finite element prediction,
- the computational complexity of the reduced order model is an increasing function of n.

As explained above, when the latent space is a vector subspace $\mathcal{V}_n$, both Galerkin projection and Céa's lemma hold, but simulation speedup may not be achieved. The study of more complex situations is the purpose of this Chapter. An estimation of computational complexity of projection-based reduced order model is proposed in Sect. 2.3.6.

Remarks:

- In the Rayleigh-Ritz method a small set of trial functions that satisfy the boundary conditions for $\widetilde{u}$ is introduce to span a solution space. This solution space is not related to any finite element model. The inclusion of the latent space in the ambient space $\mathcal{V}_h$ is essential for model order reduction.
- The finite element ambient space $\mathcal{V}_h$ incorporates homogeneous Dirichlet boundary conditions ($\widetilde{u} = 0$) on a boundary of the domain where the partial differential equations are set up). When such boundary conditions may change in the instances $(\widetilde{u}_N^{(i)})_{i=1,\ldots,m}$ these conditions must be taken into account as a linear constraint that supplement the partial differential equation. Such an issue appears when considering contact problems [38] for instance.
- Important limitations of projection-based model reduction methods includes situations where the geometry has to be handled in the exploitation phase of the reduced-order models, for instance when the problem features contact boundary conditions, crack propagation or when the geometry is a variability of the problem to learn. Geometrical variabilities are handled in the authors' works [1, 2, 22, 60, 61, 92].

2.2 High-Fidelity Model (HFM)

Consider an abstract partial differential equation in a domain Ω, with a μ-variability:

$$\mathcal{D}(\widetilde{u}; \xi, \mu) = 0, \quad \xi \in \Omega, \ \widetilde{u} \in \mathcal{V}. \tag{2.4}$$

The weak form of this partial differential equation reads: find $\widetilde{u} \in \mathcal{V}$ such that

$$\int_\Omega \widetilde{v} \, \mathcal{D}(\widetilde{u}; \xi, \mu) \, d\xi = 0, \quad \forall \widetilde{v} \in \mathcal{V}. \tag{2.5}$$

As an illustration, the concepts of this chapter are illustrated on a nonlinear structural mechanics problem, for which details on the high-fidelity model are provided in this Section. For an example with another physics, we refer to the authors' work [21], where a nonlinear transient thermal problem is considered.

The mechanical structure occupies the domain Ω, whose boundary $\partial\Omega$ is partitioned as $\partial\Omega = \partial\Omega_D \cup \partial\Omega_N$ such that $\partial\Omega_D^\circ \cap \partial\Omega_N^\circ = \emptyset$, see Fig. 2.1.

The structure is subjected to a quasi-static time-dependent loading, composed of an homogeneous Dirichlet boundary conditions on $\partial\Omega_D$ and Neumann boundary conditions on $\partial\Omega_N$ in the form of a prescribed traction T_N, as well as a volumic force f. The setting depends on some variability μ, which can be a parameter vector, or represent some nonparametrized variability. The evolution of the displacement $\widetilde{u}^\mu(\xi, t)$ over $(\xi, t) \in \Omega \times [0, T]$ is the solution of the following equations:

Fig. 2.1 Schematic
representation of the
structure of interest [18]

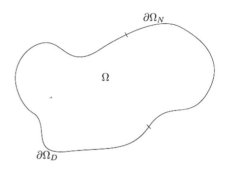

$$\epsilon(\widetilde{u}^\mu) = \frac{1}{2}\left(\nabla\widetilde{u}^\mu + (\nabla\widetilde{u}^\mu)^T\right), \qquad \text{in } \Omega \times [0, T], \quad \text{(compatibility equation)}$$

$$\text{div}\left(\sigma_c^\mu\right) + f^\mu = 0, \qquad \text{in } \Omega \times [0, T], \quad \text{(equilibrium equation)}$$

$$\sigma_c^\mu = \sigma_c(\epsilon(\widetilde{u}^\mu), y^\mu), \qquad \text{in } \Omega \times [0, T], \quad \text{(constitutive law)}$$

$$\widetilde{u}^\mu = 0, \qquad \text{in } \partial\Omega_D \times [0, T], \quad \text{(prescribed zero displacement)}$$

$$\sigma_c^\mu \cdot n_{\partial\Omega} = T_N^\mu, \qquad \text{in } \partial\Omega_N \times [0, T], \quad \text{(prescribed traction)}$$

$$\widetilde{u}^\mu = 0, \; y^\mu = 0, \qquad \text{in } \Omega \text{ at } t = 0, \quad \text{(initial condition)}$$

$$(2.6)$$

where ϵ is the linear strain tensor, σ_c^μ is the Cauchy stress tensor, y^μ denotes the internal variables of the constitutive law and $n_{\partial\Omega}$ is the outward normal vector on $\partial\Omega$. We precise that the evolution of the internal variables y^μ is updated when the constitutive law is solved.

Define $H_0^1(\Omega) = \{v \in L^2(\Omega)| \; \frac{\partial v}{\partial \xi_i} \in L^2(\Omega), \; 1 \le i \le 3 \text{ and } v|_{\partial\Omega_D} = 0\}$. Denote $\{\varphi_i\}_{1 \le i \le N} \in \mathbb{R}^{N \times N}$, a finite element basis whose span, denoted $\mathcal{V}_h$, constitutes an approximation of $H_0^1(\Omega)^3$; N is the number of finite element basis functions, hence the number of degrees of freedom of the discretized prediction. A discretized weak formulation reads: find $u^\mu \in \mathcal{V}_h$ such that for all $v \in \mathcal{V}_h$,

$$\int_\Omega \sigma_c(\epsilon(u^\mu), y) : \epsilon(v) = \int_\Omega f^\mu \cdot v + \int_{\partial\Omega_N} T_N^\mu \cdot v, \qquad (2.7)$$

that we denote for concision $\mathbf{F}^\mu(\mathbf{u}^\mu) = 0$, where $\mathbf{u}^\mu$ is the vector of N coordinates for $u^\mu \in \mathcal{V}_h$. A Newton algorithm can be used to solve this nonlinear global equilibrium problem at each time step:

$$\frac{D\mathbf{F}^\mu}{Du}\left(\mathbf{u}^{\mu,k}\right)\left(\mathbf{u}^{\mu,k+1} - \mathbf{u}^{\mu,k}\right) = -\mathbf{F}^\mu\left(\mathbf{u}^{\mu,k}\right), \qquad (2.8)$$

where

$$\frac{D\mathbf{F}^\mu}{Du}\left(\mathbf{u}^k\right)_{ij} = \int_\Omega \epsilon(\varphi_j) : \mathcal{K}\left(\epsilon(u^{\mu,k}), y^\mu\right) : \epsilon(\varphi_i), \qquad (2.9)$$

and

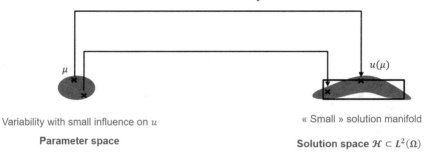

Fig. 2.2 Linear manifold learning

$$\mathbf{F}^{\mu}\left(\mathbf{u}^{\mu,k}\right)_i = \int_{\Omega} \sigma_c\left(\epsilon(u^{\mu,k}), y^{\mu}\right) : \epsilon\left(\varphi_i\right) - \int_{\Omega} f^{\mu} \cdot \varphi_i - \int_{\partial \Omega_N} T_N^{\mu} \cdot \varphi_i. \quad (2.10)$$

In the two relations above, $\mathcal{K}\left(\epsilon(u^{\mu,k}), y^{\mu}\right)$ is the local tangent operator, $u^{\mu,k} \in \mathcal{V}$ is the k-th iteration of the discretized displacement field at the current time-step, and

$$\mathbf{u}^{\mu,k} = \left(u_i^{\mu,k}\right)_{1 \leq i \leq N} \in \mathbb{R}^N \text{ is such that } u^{\mu,k} = \sum_{i=1}^{N} u_i^{\mu,k} \varphi_i. \text{ In particular, } f^{\mu}, T_N^{\mu}, u^{\mu,k}$$

and y^{μ} are known and enforce the time-dependence of the solution. Depending on the constitutive law, the computation of the functions $\left(u^{\mu,k}, y\right) \mapsto \sigma_c\left(\epsilon(u^{\mu,k}), y^{\mu}\right)$ and $\left(u^{\mu,k}, y^{\mu}\right) \mapsto \mathcal{K}\left(\epsilon(u^{\mu,k}), y^{\mu}\right)$ can require the resolution of a complex differential-algebraic system of equations.

2.3   Linear Manifold Learning for Projection-Based Reduced-Order Modeling

We start by explaining the online phase. Since we want to construct and solve the reduced-order model (ROM) in the most efficient way, the offline phase is dedicated to precompute as many steps as possible, under the considered variability.

Linear manifold learning means that the solution manifold is approximated by a vector subspace of the ambient solution space, as illustrated in Fig. 2.2.

2.3.1   Approaches Preceding the Use of Machine Learning

In structural mechanics, normal modes have been introduced for the analysis of vibrations in structures. When considering free vibrations, without external force, the solution of linear hyperbolic equation is sought by using the separation of space and time variables: $\mathbf{u}_N(\mathbf{x}, t) = \boldsymbol{\psi}(\mathbf{x}) \, \widehat{u}(t)$, where $\mathbf{x} \in \Omega$ is the space variable and $t \in \mathbb{R}$ the time variable. The hyperbolic equation of free vibration reads: find $\boldsymbol{\psi}(\mathbf{x}) \in \mathcal{V}_N$ and $\widehat{u}(t) \in \mathbb{R}$ such that $\mathbf{u}_N(\mathbf{x}, t) = \boldsymbol{\psi}(\mathbf{x}) \, \widehat{u}(t)$ and

$$\langle \rho \, \ddot{\mathbf{u}}_N, \mathbf{v}_N \rangle + a(\mathbf{u}_N, \mathbf{v}_N) = 0, \quad \forall \, \mathbf{v}_N \in \mathcal{V}_N. \tag{2.11}$$

It follows that $\widehat{u}(t)$ is an harmonic function of frequency denoted by f and $\boldsymbol{\psi}$ is the eigenvector related to the eigenvalue $\lambda = (2 \, \pi \, f)^2$ of the following generalized eigenproblem: find $\boldsymbol{\psi} \in \mathcal{V}_N$ and λ such that

$$a(\boldsymbol{\psi}, \mathbf{v}_N) - \lambda \, \langle \rho \, \boldsymbol{\psi}, \mathbf{v}_N \rangle = 0, \quad \forall \, \mathbf{v}_N \in \mathcal{V}_N, \tag{2.12}$$

where the rank of this system of equation is supposed to be $N - 1$ in order to find non zero eigenmodes. This eigenproblem admits N orthogonal normal modes that span $\mathcal{V}_N$. Therefore a selection of n normal modes span a reduced subspace of dimension n. In the beginning of the 21st century, model reduction using variable separation scheme in partial differential equations has been extended to an arbitrary number of variable by using low-rank approximations such as the Proper Generalized Decomposition [5]. Eigenmodes are global functions in contrast to finite element shape functions that have a local support. For dynamical problems involving nonlinear contributions to the PDE, adaptive computations of reduced subspaces have been proposed by Almroth et al. [4] and Noor et al. [79], by using Rayleigh-Ritz global functions. The set of these global functions is a reduced basis of the finite-element approximation-space.

The idea of using statistics to generate a solution space for differential equations has been proposed in the seminal work of Lorenz in [69] (in page 31), by using empirical orthogonal functions. The Galerkin projection of PDEs on empirical modes have been first developed in [13], where a reduced basis is computed via the proper orthogonal decomposition [70] of observational data. This was the first step towards manifold learning for projection-based model order reduction, which we now present. For other presentations of this technologies, the reader can refer to [81, 87].

2.3.2   Online Phase: Galerkin Projection

The reduced-order model is constructed in the form of a Galerkin method written on a Reduced-Order Basis (ROB). In the present case, it consists in assembling the physical problem in the same fashion as the HFM in Sect. 2.2, with the difference that the finite element basis $(\varphi_i)_{1 \leq i \leq N} \in \mathbb{R}^{N \times N}$, is replaced by a ROB $(\psi_i)_{1 \leq i \leq n} \in \mathbb{R}^{n \times N}$, with $n \ll N$. Hence, the reduced Newton algorithm is constructed as

$$\frac{D\mathcal{F}_\mu}{D\widehat{u}} \left(\widehat{u}_\mu^k \right) \left(\gamma^{\mu, k+1} - \gamma^{\mu, k} \right) = -\mathcal{F}_\mu \left(\widehat{u}_\mu^k \right), \tag{2.13}$$

where

$$\frac{D\mathcal{F}_\mu}{D\hat{u}}\left(\hat{u}_\mu^k\right)_{ij} = \int_\Omega \epsilon\left(\psi_j\right) : \mathcal{K}\left(\epsilon(\hat{u}_\mu^k), y_\mu\right) : \epsilon\left(\psi_i\right) \tag{2.14}$$

and

$$\mathcal{F}_\mu\left(\hat{u}_\mu^k\right)_i = \int_\Omega \sigma\left(\epsilon(\hat{u}_\mu^k), y\right) : \epsilon\left(\psi_i\right) - \int_\Omega f_\mu \cdot \psi_i - \int_{\partial\Omega_N} T_{\mu,N} \cdot \psi_i. \tag{2.15}$$

In the two relations above, $\hat{u}_\mu^k \in \hat{\mathcal{V}} := \mathrm{Span}\,(\psi_i)_{1 \le i \le n}$ is the k-th iteration of the reduced displacement field for the current time-step and $\boldsymbol{\gamma}^{\mu,k} = \left(\gamma_i^{\mu,k}\right)_{1 \le i \le n} \in \mathbb{R}^n$ is such that

$$\hat{u}_\mu^k = \sum_{i=1}^n \gamma_i^{\mu,k} \psi_i. \tag{2.16}$$

Notice that the use of the Galerkin method is made possible by the linearity of the tangent problem (2.13) and the choice of a linear dimensionality reduction technique in (2.16).

The online stage is called *efficient* if the reduced problem can be constructed and solved in computational complexity independent of N. When the variability μ is parametrized, efficiency is possible by precomputing various terms. With non-parametrized variability, depending on its nature, some assembling task with a linear complexity in N may be required at the beginning of the online stage (for instance for a boundary condition). All these scenarios are handled by genericROM, the ROM library developped at Safran and presented in Sect. 4.2.

The offline phase contains three steps, for which we present below the methodological choices made in genericROM.

2.3.3 Offline Phase

2.3.3.1 Data Generation

This step corresponds to the generation of the snapshots by solving the high-fidelity model. In parametric contexts, the simplest workflows consist in choosing parameter values a priori, following Design of Experiments (DoE, see [50] for a recent technique), and computing the corresponding snapshots by solving the high-fidelity model.

2.3.3.2 Data Compression: Dimensionality Reduction

This step corresponds to the generation of the ROB $(\psi_i)_{1\leq i\leq n}$. One of the most classical and simple method is the snapshot Proper Orthogonal Decomposition (POD) [25, 86], detailed below:

1. Choose a tolerance ϵ_{POD}.
2. Compute the correlation matrix $C_{i,j} = \int_\Omega u_i \cdot u_j$, $1 \leq i, j \leq N_s$, where N_s is the total number of HFM snapshots.
3. Compute the ϵ_{POD}-truncated eigendecomposition of C: $\xi_i \in \mathbb{R}^{N_c}$ and $\lambda_i > 0$, where $1 \leq i \leq n$, are the n first eigenvector and eigenvalues.
4. Compute the reduced order basis $\psi_i(x) = \dfrac{1}{\sqrt{\lambda_i}} \displaystyle\sum_{j=1}^{N_s} u_j(x)\xi_{ij}$, $1 \leq i \leq n$.

The advantages of the snapshot-POD are a reasonable computational complexity when the number of degree of freedom of the high-fidelity model are much larger than the number of snapshots, and the fact that this algorithm can be easily parallelized.

Variants can be used, for instance in the Reduced Basis [84] and the POD-greedy [41] methods, with respectively an orthonormalization of the computed high-fidelity snapshots and the incremental Singular Value Decomposition (SVD) [15].

2.3.3.3 Operator Compression

A ROM is called online-efficient if in the online stage, the reduced problems can be constructed and solved in computational complexity independent of N. The operator compression step consists in additional treatments required for the efficiency of the online stage, by pre-processing some computationally demanding integration tasks over the high-fidelity domain Ω and $\partial\Omega_N$. We notice that without any additional treatment, the numerical integration involved in the assembling of Eq. (2.13) strongly limits in practice the efficiency of the ROM: no speedup with respect to the high-fidelity model can be obtained in practice. The complexity of such an additional treatment depends on the type of parameter-dependence of the problem. This step is actually needed for all classes of problems reduced by projection-based methods.

Consider the simplest case: a linear problem with an affine dependence in the parameter μ, for instance $A_\mu \mathbf{u} = \mathbf{c}$, where $A_\mu = A_0 + \mu A_1$. Denote $\mathbf{V}$, the matrix whose columns are the vectors of the ROB evaluated at the high-fidelity degrees of freedom. The obtained ROM writes $\mathbf{V}^T A_\mu \mathbf{V}\hat{\mathbf{u}} = \mathbf{V}^T \mathbf{c}$: it is not assembled in the online phase, but rather the matrices $\mathbf{V}^T A_0 \mathbf{V}$ and $\mathbf{V}^T A_1 \mathbf{V}$ and the vector $\mathbf{V}^T \mathbf{c}$ are precomputed in the offline stage so that the reduced problem is constructed without approximation and efficiently by summing two small matrices. The operator compression step consists in this case in the construction of $\mathbf{V}^T A_0 \mathbf{V}$ and $\mathbf{V}^T A_1 \mathbf{V}$ in the offline stage.

Actually, there exist linear problems for which the operation compression step require an additional approximation, and nonlinear problems that can be carried-out exactly. In the first case, consider $A_\mu \mathbf{u} = \mathbf{c}$ with $A_{ij} = \int_\Omega \nabla \left(g(x, \mu) \varphi_j(x) \right) \cdot \nabla \varphi_i(x)$ and $c_i = \int_\Omega f(x) \varphi_i(x)$, where u is the unknown, f a known loading and $g(x, \mu)$ a known function with variables x and μ that cannot be separated: the previous precomputation of reduced matrices cannot be applied, and a treatment is required to, for example, approximately separate the dependencies in x and μ of g as $g(x, \mu) \approx \sum_{k=1}^d g_k^a(x) g_k^b(\mu)$. Then, $\mathbf{V}^T A_\mu \mathbf{V} \approx \sum_{k=1}^d g_k^b(\mu) A_k$ where $(A_k)_{ij} = \int_\Omega \nabla \left(g_k^a(x) \varphi_j(x) \right) \cdot \nabla \varphi_i(x)$, so that the efficiency of the online stage is recovered; the Empirical Interpolation Method has been proposed in [14, 71] for this purpose. A case of linear problem in harmonic aeroacoustics with nonaffine dependence with respect to the frequency is available in [20]. Conversely, nonlinearities can be handled without approximation in some cases, for instance, the advection term in fluid dynamics can be precomputed in the form of an order-3 tensor: $\int_\Omega \psi_i \cdot \left(\psi_j \cdot \nabla \right) \psi_k, 1 \leq i, j, k \leq n$; see [3] for the reduction of the nonlinear Navier-Stokes equations with an exact operator compression step. Other examples are found in structural dynamics with geometric nonlinearities, where order-2 and -3 tensors can also be precomputed, see [58, Sect. 3.2] and [73].

When additional approximations are required, the methods proposed for the operator compression step are call "hyper-reduction" in the literature. This term was coined by the seminal method proposed in [88] in 2005, but has been extended to refer to all the methods proposing a such second reduction stage. Hyper-reduction methods include the Empirical Interpolation Method (EIM, [14]), the Missing Point Estimation (MPE, [12]), the Best Point Interpolation Method (BPIM, [78]), the Discrete Empirical Interpolation Method (DEIM, [23]), the Gauss-Newton with Approximated Tensors (GNAT, [17]), the Energy-Conserving Sampling and Weighting (ECSW, [36]), the Empirical Cubature Method (ECM, [45]), and the Linear Program Empirical Quadrature Procedure (LPEQP, [100]). The reader can find an algorithmic comparison of the Hyper-Reduction and the Discrete Empirical Interpolation Method for a nonlinear thermal problem in [39]. A particular focus is given on hyper-reduction techniques via oblique projection and empirical cubature in the following sections.

2.3.4 Hyper-Reduction via a Reduced Integration Domain

Hyper-reduction via a reduced integration domain has been proposed in [88]. It requires a train set of displacement predictions, so that a reduced approximation vector space can be trained. The finite elements simulations that generate the train set of displacement fields are also predicting stresses fields σ and internal variables $\mathbf{y}$. Hence, additional reduced bases can be trained for these variables, by using these simulation results [89]. Heuristically, we found it more accurate to include a reduced basis for the stresses σ, as an additional reduced basis for this hyper-reduction scheme. Such a reduced basis is also very convenient for error estimation [91]. The finite element shape functions for displacement fields are denoted by $(\boldsymbol{\varphi}_i)_{i=1,\dots,N}$. For stress fields, we also need to introduce a related finite element representation. We denote by $(\boldsymbol{\varphi}_i^\sigma)_{i=1,\dots,N^\sigma}$ the dedicated shape functions. In the linear framework of manifold learning, we assume that the same finite element mesh is used for the target simulation, in the online step or for the test set of data, and all simulations used to generate the train set of data.

In practice, the implementation of the hyper-reduction follows the manifold learning step that trains reduced bases for displacements and stresses. We recall that they are respectively denoted by $\mathbf{V} \in \mathbb{R}^{N \times n}$ and $\mathbf{V}^\sigma \in \mathbb{R}^{N^\sigma \times n^\sigma}$ in their matrix form, and $(\boldsymbol{\psi}_k)_{k=1,\dots,n}$ $(\boldsymbol{\psi}_k^\sigma)_{k=1,\dots,n^\sigma}$ in their continuous form:

$$\boldsymbol{\psi}_k(\mathbf{x}) = \sum_{i=1}^{N} \boldsymbol{\varphi}_i(\mathbf{x}) V_{ik}, \quad \forall \mathbf{x} \in \Omega, \ k = 1, \dots, n, \tag{2.17}$$

$$\boldsymbol{\psi}_k^\sigma(\mathbf{x}) = \sum_{i=1}^{N^\sigma} \boldsymbol{\varphi}_i^\sigma(\mathbf{x}) V_{ik}^\sigma, \quad \forall \mathbf{x} \in \Omega, \ k = 1, \dots, n^\sigma. \tag{2.18}$$

The reduced displacement reads:

$$\widehat{\mathbf{u}}(\mathbf{x}) = \sum_{k=1}^{n} \boldsymbol{\psi}_k(\mathbf{x}) \gamma_k, \ \forall \mathbf{x} \in \Omega. \tag{2.19}$$

The hyper-reduction method proposed in [88] aims at computing reduced coordinates $(\gamma_k)_{k=1,\dots,n}$ introduced in Eq. (2.19), by projecting the equilibrium equation on $\mathbf{V}$, via a restriction of the domain Ω to a Reduced Integration Domain (RID) denoted by Ω_R. By following the empirical interpolation method [14], interpolation points are computed for column vectors in $\mathbf{V}$ and $\mathbf{V}^\sigma$ separately [48]. The set of respective interpolation points are denoted by $\mathcal{P}^u$ and $\mathcal{P}^\sigma$. We follow Algorithm 2.1, proposed for the Discrete Empirical Interpolation Method [23].

The RID Ω_R is such that it contains the interpolation points related to $\mathbf{V}$ and $\mathbf{V}^\sigma$. For engineering applications, the RID can also include a zone of interest in Ω that is user-defined, by using a subset of finite elements. This zone of interest is denoted by $\Omega_{ZI} \subset \Omega$. By construction, for contact-free problems, the RID is the following:

$$\Omega_R = \Omega_{ZI} \cup_{i \in \mathcal{P}^u} \text{supp}(\boldsymbol{\varphi}_i) \cup_{i \in \mathcal{P}^\sigma} \text{supp}(\boldsymbol{\varphi}_i^\sigma), \tag{2.20}$$

where $\text{supp}(f) \in \Omega$ is the support of function f. In practice, Ω_R has its own finite-element mesh. It is a reduced mesh involving much less elements than the original finite element mesh of Ω. It can be enlarge by adding a layer of connected element in the original mesh. Integration of constitutive equations are performed on this reduced mesh, without any intrusive operation on the original finite element solver. A similar hyper-reduction scheme has been developed in [38] for contact problems.

Algorithm 2.1: Interpolation points of the Discrete Empirical Interpolation Method (DEIM) [23]

Input : reduced basis vectors $\mathbf{V}[:, k] \in \mathbb{R}^N, k = 1, \dots M$
Output: interpolation point index set $I := \{i_1, \dots, i_M\}$

1 set $I_0 := \emptyset$;   // initialization
2 **for** $l = 1 \dots, M$ **do**
3     $\mathbf{U}_{l-1} \leftarrow \mathbf{V}[:, 1 : (l-1)]$;   // truncated basis
4     $\mathbf{A} \leftarrow (\mathbf{U}_{l-1}[I_{l-1}, :]^T \mathbf{U}_{l-1}[I_{l-1}, :])^{-1} \mathbf{U}_{l-1}[I_{l-1}, :]^T$;   // projector
5     $\mathbf{q}_l \leftarrow \mathbf{V}[:, l] - \mathbf{U}_{l-1}\mathbf{A}\mathbf{V}[I_{l-1}, l]$;   // interpolation residual
6     $i_l \leftarrow \arg\max_{i \in \{1, \dots, N\}} |(\mathbf{q}_l)_i|$;   // maximum of residual
7     $I_l := I_{l-1} \cup \{i_l\}$;   // extend interpolation points
8 **end**
9 set $I := I_M$.

Once the RID is obtained, a set of test functions is set up in order to restrain the balance equations to Ω_R. They are denoted by $\boldsymbol{\psi}_{R\,j}$:

$$\mathcal{P} = \left\{ i \in \{1, \dots, N\}, \int_{\Omega \setminus \Omega_R} (\boldsymbol{\varphi}_i)^2 \, d\Omega = 0 \right\}, \tag{2.21}$$

$$\boldsymbol{\psi}_{R\,j}(\mathbf{x}) = \sum_{i \in \mathcal{P}} \boldsymbol{\varphi}_i(\mathbf{x}) \, V_{ij}, \quad \forall \mathbf{x} \in \Omega, \; j = 1, \dots, n, \tag{2.22}$$

where $\mathcal{P}$ is the set of all degrees of freedom in Ω_R excepted those belonging to the interface between Ω_R and its counterpart. This interface is denoted by $\mathcal{I}_R$. As explained in [94], the test functions are null on the interface $\mathcal{I}_R$, as if Dirichlet boundary conditions were imposed to the RID. On this interface, displacements follow the shape of the modes $\boldsymbol{\psi}_k$ according to Eq. (2.19). The hyper-reduction method gives access to reduced coordinates $(\gamma_k)_{k=1,\dots,n}$ that fulfill the following balance equations, for contactless problems:

$$\widehat{\mathbf{u}}(\mathbf{x}) = \sum_{k=1}^n \boldsymbol{\psi}_k(\mathbf{x}) \gamma_k, \; \forall \mathbf{x} \in \Omega_R \tag{2.23}$$

$$\int_{\Omega_R} \boldsymbol{\varepsilon}(\boldsymbol{\psi}_{R\,j}) \; : \; \boldsymbol{\sigma}(\boldsymbol{\varepsilon}(\widehat{\mathbf{u}})) \, d\Omega \tag{2.24}$$

$$-\int_{\Omega_R} \boldsymbol{\psi}_{R\,j} \, f_\mu \, d\Omega - \int_{\partial\Omega_R \cap \partial\Omega_N} \boldsymbol{\psi}_{R\,j} \, T_\mu, N \, dS = 0. \tag{2.25}$$

$$\forall j = 1, \dots, n \tag{2.26}$$

The matrix form of the hyper-reduced balance equations reads: find $\boldsymbol{\gamma} \in \mathbb{R}^n$ such that

$$\widehat{\mathbf{u}}(\mathbf{x}) = \sum_{i=1}^{N} \varphi_i(\mathbf{x}) \, \widehat{q}_i, \; \forall \mathbf{x} \in \Omega_R, \tag{2.27}$$

$$\widehat{\mathbf{q}} = \mathbf{V} \, \boldsymbol{\gamma}, \tag{2.28}$$

$$\mathcal{F}^{HR}(\boldsymbol{\gamma}) := \mathbf{V}[\mathcal{P}, :]^T \, \mathcal{F}(\mathbf{V} \, \boldsymbol{\gamma})[\mathcal{P}], \tag{2.29}$$

$$\mathcal{F}^{HR}(\boldsymbol{\gamma}) = 0, \tag{2.30}$$

where $\mathbf{V}[\mathcal{P}, :]$ denotes a row restriction of matrix $\mathbf{V}$ to indices in $\mathcal{P}$. The reduced Newton-Raphson step reads:

$$\widehat{\mathbf{u}}^k(\mathbf{x}) = \sum_{i=1}^{N} \varphi_i(\mathbf{x}) \, \widehat{q}_i^k, \; \forall \mathbf{x} \in \Omega_R, \tag{2.31}$$

$$\widehat{\mathbf{q}}^k = \mathbf{V} \, \boldsymbol{\gamma}^k, \tag{2.32}$$

$$\mathbf{K}^{HR} := \mathbf{V}[\mathcal{P}, :]^T \, \frac{D\mathcal{F}_\mu}{D\widehat{\mathbf{u}}}(\widehat{\mathbf{u}}^{k-1})[\mathcal{P}, :] \, \mathbf{V}, \tag{2.33}$$

$$\mathbf{K}^{HR} (\boldsymbol{\gamma}^k - \boldsymbol{\gamma}^{k-1}) = -\mathbf{V}[\mathcal{P}, :]^T \, \mathcal{F}(\widehat{\mathbf{u}}^{k-1})[\mathcal{P}], \tag{2.34}$$

where the reduced stiffness matrix $\mathbf{K}^{HR}$ is computed by using solely the elements of the RID Ω_R. We assume that the matrix $\mathbf{K}^{HR}$ is full rank. This assumption is always checked in numerical solutions of hyper-reduced equations. Rank deficiency may appear when the RID construction do not account for the contribution of a reduced basis dedicated to stresses.

Once the RID is represented as a finite element mesh, this hyper-reduction scheme is intrusive solely for the linear solver involved in the Newton-Raphson step and its related convergence criterion. Nevertheless, the mesh of the RID has to include labels for the set $\mathcal{P}$ or its counterpart $\mathcal{I}_R$. This counterpart is the set of degrees of freedom connected to elements of the original mesh that are not in the reduced mesh.

Remarks:

- Here the most complex operations are indeed the computation of $\mathbf{K}^{HR}$ and the solution of the reduced linear system of equations. They respectively scale linearly with $\text{card}(\mathcal{P}) \, n^2$ and n^3. Hence n^3 has to be small enough compared to N if we consider the computational complexity for the solution of sparse linear systems in the finite element method.
- Because of the spreading nature of interpolation points, most of the time, the RID is not a compact subdomain.
- The hyper-reduced order model is a kind of submodel where the displacements at the interface $\mathcal{I}_R$ follow the shape of the modes $\boldsymbol{\psi}_k$ according to Eq. (2.19).
- Finite element corrections for displacements and stresses can be easily computed over the RID once the reduced prediction have been achieved. This scheme is termed Hybrid Hyper-Reduction in [46].
- A parallel programming of the hyper-reduction method has been proposed in [95].

- Reduced order models not only save computational time, they save computational resources including energy consumption savings as explained in [90] and memory footprint [46].

Property 1: In linear elasticity, if $\mathbf{K}^{HR}$ is full rank, the hyper-reduced balance equations are equivalent to an oblique projection of the finite element prediction $\mathbf{q} \in \mathbb{R}^N$:

$$\mathbf{\Pi}^T := \mathbf{V}[\mathcal{P}, :]^T \mathbf{K}[\mathcal{P}, :], \tag{2.35}$$

$$\widehat{\mathbf{q}} = \mathbf{V} \, (\mathbf{\Pi}^T \, \mathbf{V})^{-1} \mathbf{\Pi}^T \, \mathbf{q}, \tag{2.36}$$

$$\text{and} \quad \mathbf{\Pi}^T \, \widehat{\mathbf{q}} = \mathbf{\Pi}^T \, \mathbf{q}, \tag{2.37}$$

with $\mathbf{K} \, \mathbf{q} = \mathbf{F}$. Hence the hyper-reduced prediction of the reduced vector $\boldsymbol{\gamma}$ is a minimizer for $f(\boldsymbol{\beta})$:

$$\boldsymbol{\gamma}^\star \in \mathbb{R}^n, \ f(\boldsymbol{\gamma}^\star) = \| \mathbf{\Pi}^T \, (\mathbf{V} \, \boldsymbol{\gamma}^\star - \mathbf{q}) \|_2^2. \tag{2.38}$$

Here $\mathbf{\Pi}$ is a projector for elastic stresses in Ω_R according to the reduced test functions:

$$\sum_{i=1}^N \Pi_{ik} \, (\mathbf{V} \, \boldsymbol{\gamma} - \mathbf{q})_i = \int_{\Omega_R} \boldsymbol{\varepsilon}(\boldsymbol{\psi}_{Rk}) \, : \, (\boldsymbol{\sigma}(\widehat{\mathbf{q}}) - \boldsymbol{\sigma}(\mathbf{q})) \, d\Omega. \tag{2.39}$$

The proof is straightforward. Here, $\mathbf{K}^{HR} = \mathbf{\Pi}^T \, \mathbf{V}$. The Jacobian matrix for f reads $\mathbf{J} = \mathbf{V}^T \, \mathbf{\Pi} \, \mathbf{\Pi}^T \, \mathbf{V} = (\mathbf{K}^{HR})^T \, \mathbf{K}^{HR}$. If $\mathbf{K}^{HR}$ is full rank, then $\mathbf{J}$ is symmetric definite positive and $\mathbf{J}^{-1} = (\mathbf{K}^{HR})^{-1} \, (\mathbf{K}^{HR})^{-T}$. Then, both the minimization problem and the hyper-reduced equation have a unique solution. The solution of the minimization problem is:

$$\mathbf{q}^f = \mathbf{V} \, \mathbf{J}^{-1} \mathbf{V}^T \, \mathbf{\Pi} \, \mathbf{\Pi}^T \, \mathbf{q}, \tag{2.40}$$

$$= \mathbf{V} \, (\mathbf{K}^{HR})^{-1} \, \mathbf{\Pi}^T \, \mathbf{q}, \tag{2.41}$$

$$= \mathbf{V} \, (\mathbf{K}^{HR})^{-1} \, \mathbf{V}[\mathcal{P}, :]^T \, \mathbf{K}[\mathcal{P}, :] \, \mathbf{q}, \tag{2.42}$$

$$= \mathbf{V} \, (\mathbf{K}^{HR})^{-1} \, \mathbf{V}[\mathcal{P}, :]^T \, \mathbf{F}[\mathcal{P}], \tag{2.43}$$

$$= \widehat{\mathbf{q}}. \tag{2.44}$$

As an intermediate result, Eq. (2.42) is the oblique projection.

In linear elasticity, the Céa's lemma holds. Let us denote $\mathbf{q}^\circ \in \mathbb{R}^N$ the minimizer of the upper bound in Eq. (2.3) related to this lemma:

$$\mathbf{q}^\circ = \arg\min_{\mathbf{q}^\star \in \mathbb{R}^N} \|\widetilde{u} - \sum_{i=1}^N \boldsymbol{\varphi}_i q_i^\star\|, \tag{2.45}$$

$$\widetilde{v}^\circ = \sum_{i=1}^N \boldsymbol{\varphi}_i q_i^\circ. \tag{2.46}$$

The best projection of the minimizer $\mathbf{q}^\circ$ in the approximation space is denoted by $\boldsymbol{\gamma}_P$:

$$\boldsymbol{\gamma}^P = \text{argmin}_{\mathbf{g} \in \mathbb{R}^n} (\mathbf{q}^\circ - \mathbf{V}\,\mathbf{g})^T \mathbf{M}(\mathbf{q}^\circ - \mathbf{V}\,\mathbf{g}), \tag{2.47}$$

$$\widetilde{v}^P = \sum_{i=1}^N \boldsymbol{\varphi}_i (\mathbf{V}\,\boldsymbol{\gamma}^P)_i. \tag{2.48}$$

Let us introduce an ideal reduced basis $\mathbf{V}^\circ \in \mathbb{R}^{N \times n}$ (It assumes that n is an ideal reduced dimension) such that: $\mathbf{q}^\circ = \mathbf{V}^\circ\,\boldsymbol{\gamma}^\circ$, and $\mathbf{V}^{\circ T}\mathbf{M}\mathbf{V}^\circ = \mathbf{I}$, where $M_{ij} = \langle \boldsymbol{\varphi}_i, \boldsymbol{\varphi}_j \rangle$. Hence $\boldsymbol{\gamma}^P = \mathbf{V}^T\mathbf{M}\mathbf{V}^\circ\,\boldsymbol{\gamma}^\circ$.

Property 2: In linear elasticity, the upper bound of approximation error is increased by a Chordal distance [101] between $\mathbf{V}$ and the ideal reduced basis $\mathbf{V}^\circ$:

$$\|\widetilde{u} - \widetilde{v}^\circ\| \le \|\widetilde{u} - \widetilde{v}^P\| + \|\boldsymbol{\gamma}^\circ\|_2\, d^{Ch}(\mathbf{V}^\circ, \mathbf{V}), \tag{2.49}$$

where $d^{Ch}(\mathbf{V}^\circ, \mathbf{V})$ is the Chordal distance between $\mathbf{V}^\circ$ and $\mathbf{V}$.

Hence, the smaller the Chordal distance between the sub-space spanned by $\mathbf{V}$ and $\mathbf{V}^\circ$, the better the reduced prediction by using a Galerkin projection (When the RID covers the full domain). A certification of the reduced projection can be achieved, when all errors admit an upper bound, by following the constitutive relation error proposed in [47, 55].

The Chordal distance uses the principal angles $\boldsymbol{\theta} \in \mathbb{R}^n$, $\theta_k \in [0, \pi/2[$ for $k = 1, \ldots, n$, computed via a full singular value decomposition:

$$\mathbf{V}^T\,\mathbf{M}\,\mathbf{V}^\circ = \mathbf{U}\,\cos(\boldsymbol{\theta})\mathbf{U}^{\circ T}, \quad \mathbf{U}^T\mathbf{U} = \mathbf{U}^{\circ T}\mathbf{U}^\circ = \mathbf{I}, \tag{2.50}$$

$$d^{Ch}(\mathbf{V}^\circ, \mathbf{V}) = \|\sin(\boldsymbol{\theta})\|_F, \tag{2.51}$$

$$\|\mathbf{U}^\circ\|_F^2 = n, \tag{2.52}$$

where $\|\cdot\|_F$ is the Frobenius norm. Here, $\cos(\boldsymbol{\theta})$ and $\sin(\boldsymbol{\theta})$ are cosine and sine diagonal matrices. In addition the following property holds when a full SVD is computed:

$$\mathbf{U}^\circ\,\mathbf{U}^{\circ T} = \mathbf{I}, \quad \mathbf{U}\,\mathbf{U}^T = \mathbf{I}. \tag{2.53}$$

The proof of the previous property is straightforward by using the triangular inequality. We just need to prove that:

$$\|\widetilde{v}^\circ - \widetilde{v}^P\| \le \|\boldsymbol{\gamma}^\circ\|_2\, d^{Ch}(\mathbf{V}^\circ, \mathbf{V}).$$

Hence, the proof is the following:

$$\|\widetilde{v}^\circ - \widetilde{v}^P\|^2 = \gamma^{\circ T} \, (\mathbf{V}^\circ - \mathbf{V} \, \mathbf{V}^T \mathbf{M} \mathbf{V}^\circ)^T \, \mathbf{M} \, (\mathbf{V}^\circ - \mathbf{V} \, \mathbf{V}^T \mathbf{M} \mathbf{V}^\circ) \, \gamma^\circ$$
$$= \gamma^{\circ T} \, (\mathbf{I} - \mathbf{V}^{\circ T} \mathbf{M} \mathbf{V} \, \mathbf{V}^T \mathbf{M} \mathbf{V}^\circ) \, \gamma^\circ. \tag{2.54}$$

Therefore

$$\|\widetilde{v}^\circ - \widetilde{v}^P\|^2 = \gamma^{\circ T} \, (\mathbf{I} - \mathbf{U}^\circ \cos(\boldsymbol{\theta})^2 \, \mathbf{U}^{\circ T}) \, \gamma^\circ \tag{2.55}$$
$$= \gamma^{\circ T} \, \mathbf{U}^\circ (\mathbf{I} - \cos(\boldsymbol{\theta})^2) \, \mathbf{U}^{\circ T} \, \gamma^\circ \tag{2.56}$$
$$= \gamma^{\circ T} \, \mathbf{U}^\circ \sin(\boldsymbol{\theta})^2 \, \mathbf{U}^{\circ T} \, \gamma^\circ \tag{2.57}$$
$$= \|\sin(\boldsymbol{\theta}) \, \mathbf{U}^{\circ T} \, \gamma^\circ\|_2^2. \tag{2.58}$$

For all matrices $\mathbf{A} \in \mathbb{R}^{n \times m}$ and $\mathbf{B} \in \mathbb{R}^{m \times n}$ the following property holds:

$$\|\mathbf{AB}\|_F \leq \|\mathbf{A}\|_F \, \|\mathbf{B}\|_F,$$

and for $\mathbf{a} \in \mathbb{R}^n$: $\|\mathbf{a}\|_F = \|\mathbf{a}\|_2$.
Thus:

$$\|\widetilde{v}^\circ - \widetilde{v}^P\|^2 \leq \|\sin(\boldsymbol{\theta})\|_F^2 \, \|\mathbf{U}^{\circ T} \, \gamma^\circ\|_2^2 \leq \|\sin(\boldsymbol{\theta})\|_F^2 \, \|\gamma^\circ\|_2^2. \tag{2.59}$$

Property 3: When the identity matrix is substituted for $\mathbf{K}$ in Eqs. (2.35) and (2.36) is known as the Gappy POD reconstruction [35] of truncated variables $\mathbf{q}[\mathcal{P}]$. The reconstructed vector in $\mathbb{R}^N$ is:

$$\widetilde{\mathbf{q}} = \mathbf{V} \, (\mathbf{V}[\mathcal{P}, :]^T \, \mathbf{V}[\mathcal{P}, :])^{-1} \mathbf{V}[\mathcal{P}, :]^T \, \mathbf{q}[\mathcal{P}]. \tag{2.60}$$

This Gappy POD reconstruction is useless for displacement variables because the oblique projection in Eq. (2.36) is a direct outcome of the hyper-reduced prediction. But such a reconstruction is very convenient for stress variables that the hyper-reduced scheme forecasts only on Ω_R. The reconstructed stress variables reads:

$$\widetilde{\mathbf{q}}^\sigma = \mathbf{V}^\sigma \, (\mathbf{V}^\sigma[\overline{\mathcal{P}}^\sigma, :]^T \, \mathbf{V}^\sigma[\overline{\mathcal{P}}^\sigma, :])^{-1} \mathbf{V}^\sigma[\overline{\mathcal{P}}^\sigma, :]^T \, \mathbf{q}^\sigma[\overline{\mathcal{P}}^\sigma], \tag{2.61}$$

where $\overline{\mathcal{P}}^\sigma$ is the set of all stress indices available in Ω_R. Since the RID contains interpolation points for $\mathbf{V}^\sigma$, these points are included in $\overline{\mathcal{P}}^\sigma$, therefore the truncated matrix $\mathbf{V}^\sigma[\overline{\mathcal{P}}^\sigma, :]$ is full column rank and the reconstruction is a well posed problem.

Remark about the RID construction and the DEIM: If the RID contains solely the elements connected to interpolation points related to the reduced basis $\mathbf{V}$, such that $\mathcal{P} = \mathcal{P}^u$, then the Gappy POD gives the interpolation scheme of the DEIM:

$$\widetilde{\mathbf{q}}^{DEIM} = \mathbf{V} \, (\mathbf{V}[\mathcal{P}^u, :])^{-1} \, \mathbf{q}[\mathcal{P}^u]. \tag{2.62}$$

But, when considering the hyper-reduction scheme, one can observe overfitting in the sense that the train set of displacement is very well approximated by the DEIM reconstruction, but the hyper-reduced predictions are not accurate. For this reason, we recommend the use of the additional reduced basis $\mathbf{V}^\sigma$ and the related interpolation points.

Various applications of the hyper-reduction method using a RID have been developed for:

- thermal problems in structures or solids, in [88],
- boundary element models [93],
- reduced simulations of sintering processes [99],
- ductile damage predictions, including unstable localisation of strains [97],
- reduction of multidimensional domains, when space variables are an Euclidean space of arbitrary dimension $D > 3$ [98],
- simulation of viscoelastic-viscoplastic composites materials [74],
- model calibration in plasticity of materials [37, 46, 96],
- contact problems using Lagrange multipliers [38, 62],
- arc length algorithm for buckling problems or strain localisation [59],
- micromorphic continua including higher order stress fields [48].

2.3.5   Hyper-Reduction via Empirical Cubature

To assemble the linearized equations of the reduced Newton algorithm (2.13) when using the ROM in the online phase, hyper-reduction techniques via empirical cubature aim to compute the costly integrals over the high-fidelity domain by replacing the high-dimensional quadrature formula by a low-dimensional reduced quadrature with positive weights. The ECSW [36], ECM [45] and LPEQP [100] are methods implementing such reduced quadratures. In this section, we present the ECM, more details are available in [19].

We consider the high-fidelity model described in Sect. 2.2. The integrals involved in the assembling of the linearized Eq. (2.7) make use of high-fidelity quadrature formulas. Apply such quadrature to the reduced internal forces vector:

$$
\begin{aligned}
\hat{F}_i^{\text{int}}(t) &:= \int_\Omega \sigma\left(\epsilon(\hat{u}), y\right)(x, t) : \epsilon\left(\psi_i\right)(x) \\
&= \sum_{e \in E}\sum_{k=1}^{n_e} \omega_k \sigma\left(\epsilon(\hat{u}), y\right)(x_k, t) : \epsilon\left(\psi_i\right)(x_k),\ 1 \le i \le n,
\end{aligned}
\tag{2.63}
$$

where E denotes the set of elements of the mesh, n_e the number of quadrature points for the element e, ω_k and x_k are the quadrature weights and points associated to e. The total number of quadrature points is denoted N_G.

The ECM aims to approximate the high-fidelity quadrature by a reduced quadrature with positive weights, which, when applied to the reduced internal forces vector, writes

$$\hat{F}_i^{\text{int}}(t) \approx \sum_{k'=1}^{n_g} \hat{\omega}_{k'} \sigma \left(\epsilon(\hat{u}), y\right) (\hat{x}_{k'}, t) : \epsilon(\psi_i)(\hat{x}_{k'}), 1 \leq i \leq n, \qquad (2.64)$$

where $\hat{\omega}_{k'} > 0$ and $\hat{x}_{k'}$ are respectively the reduced quadrature weights and points, and $n_g \ll N_G$ is the length of the reduced quadrature.

Denote $f_q := \sigma \left(\epsilon(u_{(q//n)+1}), y\right) : \epsilon(\psi_{(q\%n)+1}), 1 \leq q \leq nN_c$. where $//$ and $\%$ are the quotient and the remainder of the Euclidean division. Denote as well $\mathcal{Z}^{n_G}$ a subset of $[1; N_G]$ of size n_G and $J_{\mathcal{Z}^{n_G}} \in \mathbb{R}^{nN_c \times n_G}$ and $\boldsymbol{b} \in \mathbb{N}^{nN_c}$ such that for all $1 \leq q \leq nN_c$ and all $1 \leq q' \leq n_G$,

$$J_{\mathcal{Z}^{n_G}} = \left(f_q(x_{\mathcal{Z}_{q'}^{n_G}})\right)_{1 \leq q \leq nN_c, \, q' \in \mathcal{Z}^{n_G}} \quad , \qquad \boldsymbol{b} = \left(\int_\Omega f_q\right)_{1 \leq q \leq nN_c}, \qquad (2.65)$$

where $\mathcal{Z}_{q'}^{n_G}$ denotes the q'-th element of $\mathcal{Z}^{n_G}$. We remind that n is the number of POD modes, see Sect. 2.3.3.2. Let $\hat{\boldsymbol{\omega}} \in \mathbb{R}_G^{+n}, \left(J_{\mathcal{Z}^{n_G}} \hat{\boldsymbol{\omega}}\right)_q = \sum_{q'=1}^{n_G} \hat{\omega}_{q'} \sigma \left(\epsilon(u_{(q//n)+1}), y\right)$ $(x_{\mathcal{Z}_{q'}^{n_G}}) : \epsilon(\psi_{(q\%n)+1})(x_{\mathcal{Z}_{q'}}), \; 1 \leq q \leq nN_c,$ is a candidate approximation for $\int_\Omega \sigma \left(\epsilon(u_{(q//n)+1}), y\right) : \epsilon(\psi_{(q\%n)+1}) = b_q , 1 \leq q \leq nN_c$. The problem of finding the more accurate reduced quadrature formula of length n_G for the reduced internal forces vector is:

$$\left(\hat{\omega}, \mathcal{Z}^{n_G}\right) = \arg \min_{\hat{\omega}' \in \mathbb{R}^{+n_G}, \mathcal{Z}^{m_G} \subset [1; N_G]} \left\| J_{\mathcal{Z}^{m_G}} \hat{\omega}' - b \right\|, \qquad (2.66)$$

where $\|\cdot\|$ denotes the Euclidean norm. Minimizing the length of the reduced quadrature formula as well leads to a NP-hard problem, which solution can be approximated using a Nonnegative Orthogonal Matching Pursuit algorithm, see Algorithm 2.2.

Algorithm 2.2: Nonnegative Orthogonal Matching Pursuit.

    Input : J, b, tolerance $\epsilon_{\text{Op.comp.}}$, $x_k, 1 \leq k \leq N_G$
    Output: $\hat{\omega}_k, \hat{x}_k, 1 \leq k \leq d$.
1   Set $\mathcal{Z} = \emptyset, k' = 0, \hat{\omega} = 0$ and $r_0 = b$;   // initialization
2   **while** $\|r_{k'}\|_2 > \epsilon_{\text{Op.comp.}} \|b\|_2$ **do**
3      $\mathcal{Z} \leftarrow \mathcal{Z} \cup \max \text{ index} \left(\left(J_{[1;N_G]}\right)^T r_{k'}\right)$
4      $\hat{\omega} \leftarrow \arg \min_{\hat{\omega}'>0} \left\| b - J_{\mathcal{Z}} \hat{\omega}' \right\|_2$
5      $r_{k'+1} \leftarrow b - J_{\mathcal{Z}} \hat{\omega}$
6      $k' \leftarrow k' + 1$
7   **end**
8   $d \leftarrow k'$
9   $\hat{x}_k := x_{\mathcal{Z}_k}, 1 \leq k \leq d$

In Algorithm 2.2, $J_{[1;N_G]}$ satisfies the definition (2.65) with $\mathcal{Z}^{n_G} = [1; N_G]$. The positivity of the weights of the reduced quadrature preserves the spectral properties of the operator associated with the high-fidelity problem, see [19, Remark 1].

2.3.6 Computational Complexity

In this section we restrict our attention to elliptic problems or to linearized problems. The bilinear part of the weak form for finite-dimensional solution spaces is a matrix. When using Finite Element solution space, this matrix is sparse. But when using a reduced solution space, this matrix is usually a full matrix. Therefore, computational complexity of the finite element prediction is the complexity of the solution of sparse linear system. It scales linearly with $N \omega^2$, where ω is the band width of the sparse matrix. For the reduced prediction, the solution of a full linear system scales linearly with n^3. We recommend to restrict linear model reduction schemes, with or without hyper-reduction, to reduced dimension n lower than $N^{1/3}$, otherwise the solution of reduced equation will have a computational complexity similar to the complexity of the finite element model. This recommendation does not concern explicit solvers.

2.4 Nonlinear Manifold Learning for Projection-Based Reduced-Order Modeling

Consider a parametrized variability, and a set of snapshots generated using the high-fidelity model over a sampling of the parameter domain. The parametrized problem is said nonreducible when applying a linear data compression over this set of snapshots leads to a ROB containing too many vectors for the online problem to feature an interesting speedup. Formally, this happens when the Kolmogorov n-width $d_n(\mathcal{M})$ decreases too slowly with respect to n, where we recall that n is the cardinality of the ROB,

$$d_n(\mathcal{M}) := \inf_{\mathcal{H}_n \in \mathrm{Gr}(n,\mathcal{H})} \sup_{u \in \mathcal{M}} \inf_{v \in \mathcal{H}_n} \|u - v\|_{\mathcal{H}}, \tag{2.67}$$

with the Grassmannian $\mathrm{Gr}(n, \mathcal{H})$ being the set of all n-dimensional subspaces of $\mathcal{H}$ and $\mathcal{H}_n \in \mathrm{Gr}(n, \mathcal{H})$ the subspace spanned by the considered ROB. Qualitatively, the solution manifold $\mathcal{M}$ covers too many independent directions to be embedded in a low-dimensional subspace. To address this issue, several techniques have been developed:

- Problem-specific methods tackle the difficulties of some specific physics problems that are known to be nonreducible, such as advection-dominated problems which have been largely investigated, for instance in [16, 49, 85].

- Online-adaptive model reduction methods update the ROM in the exploitation phase by collecting new information online as explained in [102], in order to limit extrapolation errors when solving the parametrized governing equations in a region of the parameter space that was not explored in the training phase. The ROM can be updated for example by querying the high-fidelity model when necessary for basis enrichment [18, 44, 56, 80, 88].
- ROM interpolation methods [6–9, 24, 64–68, 75, 76] use interpolation techniques on Grassmann manifolds or matrix manifolds to adapt the ROM to the parameters considered in the exploitation phase by interpolating between two precomputed ROMs.
- Dictionaries of basis vector candidates enable building a parameter-adapted ROM in the exploitation phase by selecting a few basis vectors. This technique is presented in [54, 72] for the Reduced Basis method.
- Nonlinear manifold ROM methods [57, 63] learn a nonlinear embedding and project the governing equations onto the corresponding approximation manifold, by means of a nonlinear function mapping a low-dimensional latent space to the solution space.
- Dictionaries of ROMs rely on the construction of several local ROMs adapted to different regions of the solution manifold. These local ROMs can be obtained by partitioning the time interval [32, 33], the parameter space [33, 34, 42, 44, 51, 52, 82], or the solution space [10, 11, 27, 40, 77, 82, 92].

In the following Sects. 2.4.1 and 2.4.2, we provide more details on the last two entries of the previous list.

2.4.1 Nonlinear Dimensionality Reduction via Auto-Encoder

Nonlinear manifold learning means that the solution manifold is approximated by a domain in the ambient solution space that is not included in a low-dimensional vector subspace, as illustrated in Fig. 2.3.

Let us consider a formal representation of parabolic and nonlinear Partial Differential Equations (PDEs) that are parameterized with respect to some physical parameters of interest. We mean by physical parameters the parameters that appear directly within the equations such as the boundary conditions, the viscosity for fluid mechanics (henceforth the Reynolds number), the time step for dynamical systems of fluid flows or infectious diseases, etc. These parameters are denoted μ without any loss of generality as introduced in the preceding section. The formal representation of the equations is given as follows:

$$\frac{\partial \widetilde{u}}{\partial t} = f(\widetilde{u}, \mu). \tag{2.68}$$

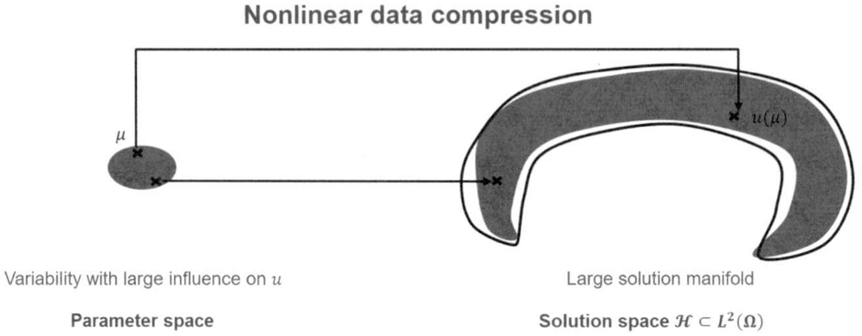

Nonlinear data compression

Variability with large influence on u Large solution manifold

Parameter space   **Solution space** $\mathcal{H} \subset L^2(\Omega)$

Fig. 2.3 Nonlinear manifold learning

Reduced order modeling based on nonlinear data compression techniques might be a solution for example when the described physical fields by the model equations require a large number of vectors in the ROM as specified above. Nevertheless there are cases where even if the physical solution fields are completely reducible, the Galerkin projection may not be appropriate for the model equations describing this physics.

Convection-diffusion PDEs have this stability issue even more generally when considering Galerkin projection using finite element basis functions. In the literature, it is proved that the coherent structures of a turbulent, unsteady and in-compressible fluid flow are reproducible by a small number of POD basis functions. However, if these functions are used to solve the Newton-Raphson problem of the associated reduced order Galerkin dynamical system, then an instability appears as a function of the time. In the literature, many solutions are proposed to tackle this difficulty while keeping the reduced order approximation in a linear space spanned by the POD basis functions. We can refer to the Petrov-Galerkin technique, the least square minimization of the equations residual, the variational finite element method, etc. Recently, nonlinear approximations of the solution fields in a manifold of reduced dimension start to gain importance in the literature. In this case, the reduced order model is said to be a nonlinear projection based reduced model. Some authors introduced nonlinear approximations using Deep Learning approaches for projection based reduced models. We find in the literature more classical nonlinear approximations based on the Kernel POD technique.

In what follows, we make a focus from the literature on Deep Learning projection based reduced models and their different possible formulations. More precisely, we are talking about Deep AutoEncoders (DAEs) from the domain of Deep Learning. DAEs are artificial neural networks formed of layers of spatial convolutions, nonlinear activation functions and linear systems called fully connected functions. These architectures are used to perform nonlinear dimensional reduction, following unsupervised data compression. Henceforth, the DAE allows to determine latent features within a set of given inputs.

We denote by h and g respectively the encoder mapping and the decoder mapping of a DAE. In general g is the transpose mapping of h. We denote by $\widehat{\alpha}$ the reduced latent features inferred by h. The dimension of the latent features is equal to the intrinsic dimensionality of the manifold as stated in Remark 2.1 in [63]. This intrinsic dimensionality is in the current case the dimension N_μ of the vector of parameters μ, which may include the time variable also.

We note that the reduced latent features of a DAE are not parameterized variables in general. In other words they can be seen as non-parametric features associated with a given set of inputs. This formulation is interesting in the framework of projection-based model reduction, where the associated parameters are given straightforwardly by the physical equations. Hence, knowing the variable parameters within the inputs data helps only with the determination of the intrinsic dimension of the manifold.

In the literature, we find two different formulations for DAEs projection based reduced models.

The first formulation was proposed by Kashima [53] and Hartman et al. [43]. It is very analogical to the Galerkin formulation of projection based reduced models: given h and g such that $g \circ h : \widetilde{u} \to \widehat{u}$ and $\widehat{u} \approx \widetilde{u}$, then the reduced model is formulated as follows:

$$\frac{\partial}{\partial t}\widehat{\alpha}(\mu) = h \circ f \circ g(\widehat{\alpha}(\mu)) \quad , \quad \widehat{\alpha}(t = 0) = h(\widetilde{u}(t = 0)) \tag{2.69}$$

$$= h\left(f(g(\widehat{\alpha}(\mu)), \mu)\right). \tag{2.70}$$

In the above formulation, the authors relied on the following three points in order to set the time derivative of the latent features equal to the right hand side of Eq. (2.69).

- $\dfrac{\partial}{\partial t}g(\widehat{\alpha}(\mu))$ belongs to the manifold described by $\mu \to \widetilde{u}(\mu)$,

- $h\left(\dfrac{\partial}{\partial t}g(\widehat{\alpha}(\mu))\right) = \dfrac{\partial}{\partial t}h\left(g(\widehat{\alpha}(\mu))\right)$,

- $h \circ g = I_{N_\mu}$.

Remark 2.1 The first two above items are hypothesis that are fulfilled in the case where h and g are linear or affine functions. The last item is fulfilled theoretically by the inputs data compression using parameters optimisation of the DAE architecture.

The second formulation of DAEs projection based reduced models is proposed in [63], where a least square minimisation of the residual of parabolic PDEs because of the decoder approximation is performed. Then, the reduced model is formulated as follows in order to determine the reduced latent features:

$$\frac{\partial}{\partial t}\widehat{\alpha}(\mu) = \operatorname{argmin}_{\widehat{v}(\mu)\in\mathbb{R}^{N_\mu}} \|J(\widehat{a}(\mu))\widehat{v}(\mu) - f(g(\widehat{a}(\mu)), \mu)\|_2^2, \tag{2.71}$$

where $\widehat{v}(\mu)$ is the time derivative of $\widehat{a}(\mu)$, $\|.\|_2$ denotes the mean square norm or the euclidean norm and, J is the Jacobian matrix of the decoder mapping which belong element-wise to the tangent space to the solutions manifold at a given point. J is expressed as follows:

$$J : \widehat{a} \rightarrow \nabla g(\widehat{a}).$$

In this second formulation, the authors do not suppose that the velocity of the decoder approximation is in the manifold of the solutions because mathematically it belongs to the tangent space to the manifold at a given point. Hence, they claim that encoding the decoder approximation will produce a poor approximation by the reduced model.

2.4.2 Piecewise Linear Dimensionality Reduction via Dictionary-Based ROM-Nets

Parts of this section has been inspired from the authors previous work [30].

Piecewise linear manifold learning means that the solution manifold is approximated by a dictionary of local linear subspaces, as illustrated in Fig. 2.4, where we denote $\mathcal{M}$ the solution manifold.

The solution manifold is partitioned to get a collection of subsets $\mathcal{M}_k \subset \mathcal{M}$ that can be covered by a dictionary of low-dimensional subspaces, enabling the use of local linear ROMs. If $\{\mathcal{M}_k\}_{k\in[\![1;K]\!]}$ is a partition of $\mathcal{M}$, then:

$$\forall k \in [\![1; K]\!], \ \forall N \in \mathbb{N}^*, \quad d_N(\mathcal{M}_k) \leq d_N(\mathcal{M}). \tag{2.72}$$

The concepts of ROM-net and dictionary-based ROM-net are introduced in [27], which we present in this section. Suppose we dispose of an already computed dictionary of ROMs for the parametrized problem (2.4), where each element of the dictionary is a ROM that can approximate the problem on a subset of the solution manifold $\mathcal{M}$. A dictionary-based ROM-net is a machine learning algorithm trained to assign the parameter $\mu \in \mathcal{X}$ to the ROM of the dictionary leading to the most accu-

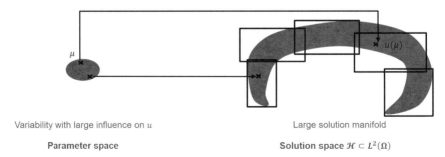

Fig. 2.4 Piecewise linear manifold learning

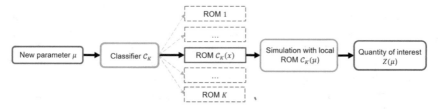

Fig. 2.5 Exploitation phase of a dictionary-based ROM-net. K local ROMs, combined with a classifier C_K for automatic ROM $C_K(\mu)$ recommendation, are used to predict the quantity of interest $Z(\mu)$

rate reduced prediction. This assignment, called model recommendation in [77], is a classification task, see Fig. 2.5.

The dictionary of ROMs is constructed in a clustering stage, during which snapshots are regrouped depending on their respective proximity on $\mathcal{M}$, in the sense of a particular dissimilarity measure we introduced in [29] and [26]. The dissimilarity between two parameter values $\mu, \mu' \in \mathcal{X}$, denoted by $\delta(\mu, \mu')$, involves the sine of the principal angles between subspaces associated to the solutions of the HFM $u(\mu), u(\mu') \in \mathcal{M}$, see [26, Definition 4.10]. Applying a k-medoids clustering algorithm on the solution manifold $\mathcal{M}$ using the dissimilarity δ leads to an optimal partitioning for a dictionary of local ROMs, in a sense introduced in [26, Property 4.13]. We refer to the remaining of [26] for the description of a practical efficiency criterion of the dictionary-based ROM-net, which enables to decide, before the computationally costly steps of the workflow, if a dictionary of ROMs is preferable to one global ROM, and how to calibrate the various hyperparameters of the ROM-net.

Remark 2.2 Importance of the classification. One could argue that the classification step can be replaced by choosing the cluster k for which the dissimilarity measure $\delta(\mu, \tilde{\mu}_k)$ between the parameter μ and the cluster medoid $\tilde{\mu}_k$ is the smallest. However, we recall that the computation of the dissimilarity measure requires solving the HFM at the parameter value μ, which would render the complete model reduction framework useless. Hence, the classification step enables to bypass this HFM solve and directly recommend the appropriate local ROM.

As briefly mentioned in the introduction of Sect. 2.4, local ROMs can be constructed by partitioning the parameter space [33, 34], in which case the classification step is not required: the cluster affectation is made by computing distances directly in the parameter space. In other cases, partitioning in the solution space can be considered without requiring a classification step [10]. Consider a time-dependent problems where the initial condition is not a parameter of the problem, and suppose an efficient computation of the clustering distance in the solution space based on the reduced solution at the previous time-step. Then, local basis affectation and switching is possible without requiring classification.

The training of the classifier can be difficult when working with physical fields: simulations are costly, data are in high dimension and classical data augmentation techniques for images cannot be applied. Hence, we can consider replacing the HFM by an intermediate-fidelity solver for generating the data needed for the training of the classifier, by considering coarser meshes and fewer time steps. We point out that the HFM should be used at the end for generating the data required in the training of the local ROMs. We propose in [28] improvements for the training of the classifier in our context by developing a fast variant of the mRMR [83] feature selection algorithm, and new class-conserving transformations of our data, acting like a data augmentation procedure.

2.5 Iterative and Greedy Strategies

For the sake of the presentation, we have separated the offline and online phases, where the Reduced-Order Model is learnt, then exploited. Actually, more involved strategies exist, where the ROM is constructed in a iterative fashion. The Reduced Basis Method [84] is a greedy method, where the ROB is constructed by a single snapshot, corresponding to a randomly chosen parameter value, and the ROB is enriched by the parameter value that maximizes the error made by the current ROM. In complex parameter dependencies, the hyper-reduction scheme can be simulateously constructed as the ROB grows, see [31] for such a scheme, with the EIM as hyper-reduction. This greedy construction as been extended to time-dependant problems in the POD-greedy method [41], and simultaneous hyper-reduction construction have also been proposed [88].

Such iterative strategies rely on a efficient computation of the error made by the ROM. Error estimation is investigated in the next chapter.

References

1. N. Akkari, F. Casenave, E. Hachem, D. Ryckelynck, A bayesian nonlinear reduced order modeling using variational autoencoders. Fluids 7(10), (2022)
2. N. Akkari, F. Casenave, D. Ryckelynck, C. Rey, An updated Gappy-POD to capture non-parameterized geometrical variation in fluid dynamics problems. Adv. Model. Simul. Eng. Sci. 9(1), 1–34 (2022)
3. N. Akkari, R. Mercier, G. Lartigue, V Moureau, Stable pod-galerkin reduced order models for unsteady turbulent incompressible flows, in 55th AIAA Aerospace Sciences Meeting, Grapevine, Texas (2017)
4. B.O. Almroth, P. Stern, F.A. Brogan, Automatic choice of global shape functions in structural analysis. AIAA J. 16(5), 525–528 (1978)
5. A. Ammar, B. Mokdad, F. Chinesta, R. Keunings, A new family of solvers for some classes of multidimensional partial differential equations encountered in kinetic theory modeling of complex fluids. J. Non Newton. Fluid Mech. 139(3), 153–176 (2006)
6. D. Amsallem, J. Cortial, C. Farhat, Towards real-time computational-fluid-dynamics-based aeroelastic computations using a database of reduced-order information. AIAA J. 48(9), 2029–2037 (2010)

7. D. Amsallem, C. Farhat, Interpolation method for adapting reduced-order models and application to aeroelasticity. AIAA J. **46**(7), 1803–1813 (2008)
8. D. Amsallem, C. Farhat, An online method for interpolating linear parametric reduced-order models. SIAM J. Sci. Comput. **33**(5), 2169–2198 (2011)
9. D. Amsallem, R. Tezaur, C. Farhat, Real-time solution of linear computational problems using databases of parametric reduced-order models with arbitrary underlying meshes. J. Comput. Phys. **326**, 373–397 (2016)
10. D. Amsallem, M. Zahr, C. Farhat, Nonlinear model order reduction based on local reduced-order bases. Int. J. Numer. Methods Eng. 1–31 (2012)
11. D. Amsallem, M. Zahr, K. Washabaugh, Fast local reduced basis updates for the efficient reduction of nonlinear systems with hyper-reduction. Adv. Comput. Math. **41**, 02 (2015)
12. P. Astrid, S. Weiland, K. Willcox, T. Backx, Missing point estimation in models described by proper orthogonal decomposition. Proc. IEEE Conf. Decis. Control. **53**(10), 1767–1772 (2005)
13. N. Aubry, P. Holmes, J.L. Lumley, E. Stone, The dynamics of coherent structures in the wall region of a turbulent boundary-layer. J. Fluid Mech. **192**, 115–173 (1988)
14. M. Barrault, Y. Maday, N.C. Nguyen, A.T. Patera, An empirical interpolation method: application to efficient reduced-basis discretization of partial differential equations. Comptes Rendus Mathematiques **339**(9), 666–672 (2004)
15. M. Brand, Incremental singular value decomposition of uncertain data with missing values, in *Computer Vision — ECCV 2002*, ed. by A. Heyden, G. Sparr, M. Nielsen, P. Johansen (Springer, Berlin, Heidelberg, 2002), pp. 707–720
16. N. Cagniart, Y. Maday, B. Stamm, Model order reduction for problems with large convection effects, in *Contributions to Partial Differential Equations and Applications*, ed. by B. Chetverushkin, W. Fitzgibbon, Y. Kuznetsov, P. Neittaanmäki, J. Periaux, O. Pironneau. Computational Methods in Applied Sciences, vol. 47 (2019)
17. K. Carlberg, F. Charbel, J. Cortial, D. Amsallem, The gnat method for nonlinear model reduction: effective implementation and application to computational fluid dynamics and turbulent flows. J. Comput. Phys. **242**, 623–647 (2013)
18. F. Casenave, N. Akkari, An error indicator-based adaptive reduced order model for nonlinear structural mechanics—application to high-pressure turbine blades. Math. Comput. Appl. **24**(2), (2019)
19. F. Casenave, N. Akkari, F. Bordeu, C. Rey, D. Ryckelynck, A nonintrusive distributed reduced-order modeling framework for nonlinear structural mechanics-application to elastoviscoplastic computations. Int. J. Numer. Methods Eng. **121**(1), 32–53 (2020)
20. F. Casenave, A. Ern, T. Lelièvre, A nonintrusive reduced basis method applied to aeroacoustic simulations. Adv. Comput. Math. **41**(5), 961–986 (2015)
21. F. Casenave, A. Gariah, C. Rey, F. Feyel, A nonintrusive reduced order model for nonlinear transient thermal problems with nonparametrized variability. Adv. Model. Simul. Eng. Sci. **7**(1), 1–19 (2020)
22. F. Casenave, B. Staber, X. Roynard, MMGP: a Mesh Morphing Gaussian Process-based machine learning method for regression of physical problems under non-parameterized geometrical variability (2023)
23. S. Chaturantabut, D. Sorensen, Discrete empirical interpolation for nonlinear model reduction, in *Proceedings of the 48th IEEE Conference on Decision and Control, 2009 held jointly with the 2009 28th Chinese Control Conference, CDC/CCC 2009*, pp. 4316–4321 (2010)
24. Y. Choi, G. Boncoraglio, S. Anderson, D. Amsallem, C. Farhat, Gradient-based constrained optimization using a database of linear reduced-order models. J. Comput. Phys. **423**, 109787 (2020)
25. L. Cordier, M. Bergmann, Proper Orthogonal Decomposition: an overview, in *Lecture series 2002-04, 2003-03 and 2008-01 on post-processing of experimental and numerical data*, Von Karman Institute for Fluid Dynamics (2008)
26. T. Daniel, F. Casenave, N. Akkari, A. Ketata, D. Ryckelynck, Physics-informed cluster analysis and a priori efficiency criterion for the construction of local reduced-order bases. J. Comput. Phys. **458**, 111120 (2022)

27. T. Daniel, F. Casenave, N. Akkari, D. Ryckelynck, Model order reduction assisted by deep neural networks (ROM-net). Adv. Model. Simul. Eng. Sci. **7**(16), (2020)
28. T. Daniel, F. Casenave, N. Akkari, D. Ryckelynck, Data augmentation and feature selection for automatic model recommendation in computational physics. Math. Comput. Appl. **26**(1), (2021)
29. T. Daniel, F. Casenave, N. Akkari, D. Ryckelynck, Optimal piecewise linear data compression for solutions of parametrized partial differential equations (2021)
30. T. Daniel, F. Casenave, N. Akkari, D. Ryckelynck, C. Rey, Uncertainty quantification for industrial numerical simulation using dictionaries of reduced order models. Mech. & Ind. **23**, 3 (2022)
31. C. Daversin, C. Prud'Homme, Simultaneous empirical interpolation and reduced basis method for non-linear problems. Comptes Rendus Mathematique **353**(12), 1105–1109 (2015)
32. M. Dihlmann, M. Drohmann, B. Haasdonk, Model reduction of parametrized evolution problems using the reduced basis method with adaptive time partitioning, vol. 01 (2011)
33. M. Drohmann, B. Haasdonk, M. Ohlberger, Adaptive reduced basis methods for nonlinear convection-diffusion equations, vol. 4, pp. 369–377, 12 (2010)
34. J. Eftang, A. Patera, E. Ronquist, An "hp" certified reduced basis method for parametrized elliptic partial differential equations. SIAM J. Sci. Comput. **32**, 3170–3200, 09 (2010)
35. R. Everson, L. Sirovich, Karhunen-Loève procedure for gappy data. J. Opt. Soc. Am. A **12**(8), 1657–1664 (1995)
36. C. Farhat, P. Avery, T. Chapman, J. Cortial, Dimensional reduction of nonlinear finite element dynamic models with finite rotations and energy-based mesh sampling and weighting for computational efficiency. Int. J. Numer. Methods Eng. **98**(9), 625–662 (2014)
37. H. Farooq, D. Ryckelynck, S. Forest et al., A pruning algorithm preserving modeling capabilities for polycrystalline data. Comput. Mech **68**, 1407–1419 (2021)
38. J. Fauque, I. Ramiere, D. Ryckelynck, Hybrid hyper-reduced modeling for contact mechanics problems. Int. J. Numer. Methods Eng. **115**(1), 117–139 (2018)
39. F. Fritzen, B. Haasdonk, D. Ryckelynck, S. Schöps, An algorithmic comparison of the hyper-reduction and the discrete empirical interpolation method for a nonlinear thermal problem. Math. Comput. Appl. **23**(1), (2018)
40. S. Grimberg, C. Farhat, R. Tezaur, C. Bou-Mosleh, Mesh sampling and weighting for the hyperreduction of nonlinear Petrov-Galerkin reduced-order models with local reduced-order bases, vol. 08 (2020)
41. B. Haasdonk, Convergence rates of the pod-greedy method. ESAIM Math. Model. Numer. Anal. **47**(3), 859–873 (2013)
42. B. Haasdonk, M. Dihlmann, M. Ohlberger, A training set and multiple bases generation approach for parametrized model reduction based on adaptive grids in parameter space. Math. Comput. Model. Dyn. Syst. **17**, 423–442, 08 (2011)
43. D. Hartman, L.K. Mestha, A deep learning framework for model reduction of dynamical systems, in *2017 IEEE Conference on Control Technology and Applications (CCTA)* (IEEE, 2017), pp. 1917–1922
44. W. He, P. Avery, C. Farhat, In-situ adaptive reduction of nonlinear multiscale structural dynamics models (2020)
45. J.A. Hernandez, M.A. Caicedo, A. Ferrer, Dimensional hyper-reduction of nonlinear finite element models via empirical cubature. Comput. Methods Appl. Mech. Eng. **313**, 687–722 (2017)
46. W. Hilth, D. Ryckelynck, C. Menet, Data pruning of tomographic data for the calibration of strain localization models. Math. Comput. Appl. **24**(1), (2019)
47. K.C. Hoang, P. Kerfriden, S.P.A. Bordas, A fast, certified and tuning free two-field reduced basis method for the metamodelling of affinely-parametrised elasticity problems. Comput. Methods Appl. Mech. Eng. **298**, 121–158 (2016)
48. M. Horák, D. Ryckelynck, S. Forest, Hyper-reduction of generalized continua. Comput. Mech **59**, 753–778 (2017)

49. A. Iollo, D. Lombardi, Advection modes by optimal mass transfer. Phys. Rev. E **89**, 022923 (2014)
50. V.R. Joseph, E. Gul, S. Ba, Maximum projection designs for computer experiments. Biometrika **102**(2), 3 (2015)
51. M.G. Kapteyn, D.J. Knezevic, K.E. Willcox, Toward predictive digital twins via component-based reduced-order models and interpretable machine learning (2020)
52. M.G. Kapteyn, K.E. Willcox, From physics-based models to predictive digital twins via interpretable machine learning (2020)
53. K. Kashima, Nonlinear model reduction by deep autoencoder of noise response data, in *2016 IEEE 55th Conference on Decision and Control (CDC)* (IEEE, 2016), pp. 5750–5755
54. S. Kaulmann, B. Haasdonk, Online greedy reduced basis construction using dictionaries (2012)
55. P. Kerfriden, J.J. Rodenas, S.P.-A. Bordas, Certification of projection-based reduced order modelling in computational homogenisation by the constitutive relation error. Int. J. Numer. Methods Eng. **97**(6), 395–422 (2014)
56. T. Kim, D.L. James, Skipping steps in deformable simulation with online model reduction. ACM Trans. Graph. **28**(5), 1–9 (2009)
57. Y. Kim, Y. Choi, D. Widemann, T. Zohdi, A fast and accurate physics-informed neural network reduced order model with shallow masked autoencoder (2020)
58. R.J. Kuether, Nonlinear modal substructuring of geometrically nonlinear finite element models. Ph.D. thesis, The University of Wisconsin-Madison, 2014
59. H. Launay, J. Besson, D. Ryckelynck, F. Willot, Hyper-reduced arc-length algorithm for stability analysis in elastoplasticity. Int. J. Solids Struct. **208–209**, 167–180 (2021)
60. H. Launay, D. Ryckelynck, L. Lacourt, J. Besson, A. Mondon, F. Willot, Deep multimodal autoencoder for crack criticality assessment. Int. J. Numer. Methods Eng. **123**(6), 1456–1480 (2022)
61. H. Launay, F. Willot, D. Ryckelynck, J. Besson, Mechanical assessment of defects in welded joints: morphological classification and data augmentation. J. Math. Ind. **11**(8), 18 (2021)
62. S. Le Berre, I. Ramière, J. Fauque, D. Ryckelynck, Condition number and clustering-based efficiency improvement of reduced-order solvers for contact problems using Lagrange multipliers. Mathematics **10**(9), (2022)
63. K. Lee, K.T. Carlberg, Model reduction of dynamical systems on nonlinear manifolds using deep convolutional autoencoders. J. Comput. Phys. **404**, 108973 (2020)
64. T. Lieu, C. Farhat, Adaptation of POD-based aeroelastic ROMs for varying Mach number and angle of attack: application to a complete F-16 configuration, in *AIAA Paper 2005-7666*, (2005)
65. T. Lieu, C. Farhat, Adaptation of aeroelastic reduced-order models and application to an F-16 configuration. AIAA J. **45**, 1244–1257 (2007)
66. T. Lieu, C. Farhat, M. Lesoinne, POD-based aeroelastic analysis of a complete F-16 configuration: ROM adaptation and demonstration, in *AIAA Paper 2005-2295* (2005)
67. T. Lieu, C. Farhat, M. Lesoinne, Reduced-order fluid/structure modeling of a complete aircraft configuration. Comput. Methods Appl. Mech. Eng. **195**, 5730–5742 (2006)
68. T. Lieu, M. Lesoinne, Parameter adaptation of reduced order models for three-dimensional flutter analysis, in *AIAA Paper 2004-0888* (2004)
69. E.N. Lorenz, Empirical orthogonal functions and statistical weather prediction. *MIT, Department of Meteorology, Scientific Report N1, Statistical Forecasting Project* (1956)
70. J.L. Lumley, The structure of inhomogeneous turbulence, in *Atmospheric Turbulence and Wave Propagation*, pp. 166–177 (1967)
71. Y. Maday, N.-C. Nguyen, A.T. Patera, S.H. Pau, A general multipurpose interpolation procedure: the magic points. Commun. Pure Appl. Anal. **8**(1), 383–404 (2009)
72. Y. Maday, B. Stamm, Locally adaptive greedy approximations for anisotropic parameter reduced basis spaces. SIAM J. Sci. Comput. **35**(6), A2417–A2441 (2013)
73. M.P. Mignolet, A. Przekop, S.A. Rizzi, S.M. Spottswood, A review of indirect/non-intrusive reduced order modeling of nonlinear geometric structures. J. Sound Vib. **332**(10), 2437–2460 (2013)

74. B. Miled, D. Ryckelynck, S. Cantournet, A priori hyper-reduction method for coupled viscoelastic-viscoplastic composites. Comput. & Struct. **119**, 95–103 (2013)
75. R. Mosquera, A. El Hamidi, A. Hamdouni, A. Falaize, Generalization of the Neville-Aitken interpolation algorithm on Grassmann manifolds : applications to reduced order model (2019). arXiv:1907.02831
76. R. Mosquera, A. Hamdouni, A. El Hamidi, C. Allery, POD basis interpolation via inverse distance weighting on Grassmann manifolds. Discret. Contin. Dyn. Syst. S **12**, 1743–1759, 01 (2018)
77. F. Nguyen, S.M. Barhli, D.P. Muñoz, D. Ryckelynck, Computer vision with error estimation for reduced order modeling of macroscopic mechanical tests. Complexity (2018)
78. N.C. Nguyen, A.T. Patera, J. Peraire, A best points interpolation method for efficient approximation of parametrized functions. Int. J. Numer. Methods Eng. **73**, 521–543 (2008)
79. A.K. Noor, J.M. Peters, Reduced basis technique for nonlinear analysis of structures. AIAA J. **18**(4), 455–462 (1980)
80. M. Ohlberger, F. Schindler, Error control for the localized reduced basis multiscale method with adaptive on-line enrichment. SIAM J. Sci. Comput. **37**(6), A2865–A2895 (2015)
81. A.T. Patera, G. Rozza, Reduced basis approximation and a posteriori error estimation for parametrized partial differential equations. MIT Pappalardo Graduate Monographs in Mechanical Engineering (2007)
82. B. Peherstorfer, D. Butnaru, K. Willcox, H.J. Bungartz, Localized discrete empirical interpolation method. SIAM J. Sci. Comput. **36**, 01 (2014)
83. H. Peng, F. Long, C. Ding, Feature selection based on mutual information: Criteria of max-dependency, max-relevance, and min-redundancy. IEEE Trans. Pattern Anal. Mach. Intell. **27**, 1226–38, 09 (2005)
84. C. Prud'Homme, D.V. Rovas, K. Veroy, L. Machiels, Y. Maday, A.T. Patera, G. Turinici, Reliable real-time solution of parametrized partial differential equations: reduced-basis output bound methods. J. Fluids Eng. **124**(1), 70–80 (2001)
85. J. Reiss, P. Schulze, J. Sesterhenn, V. Mehrmann, The shifted proper orthogonal decomposition: a mode decomposition for multiple transport phenomena. SIAM J. Sci. Comput. **40**(3), A1322–A1344 (2018)
86. C. Rowley, T. Colonius, R. Murray, Model reduction for compressible flow using POD and Galerkin projection. Phys. D Nonlinear Phenom. **189**, 115–129, 01 (2003)
87. G. Rozza, D. Huynh, A. Patera, Reduced basis approximation and a posteriori error estimation for affinely parametrized elliptic coercive partial differential equations. Arch. Comput. Methods Eng. **15**, 1–47, 09 (2007)
88. D. Ryckelynck, A priori hyperreduction method: an adaptive approach. J. Comput. Phys. **1**(202), 346–366 (2005)
89. D. Ryckelynck, Hyper reduction of mechanical models involving internal variables. Int. J. Numer. Methods Eng. (2009)
90. D. Ryckelynck, D.M. Benziane, A. Musienko, G. Cailletaud, Toward "green" mechanical simulations in materials science. Eur. J. Comput. Mech. **19**(4), 365–388 (2010)
91. D. Ryckelynck, L. Gallimard, S Jules, Estimation of the validity domain of hyper-reduction approximations in generalized standard elastoviscoplasticity. Adv. Model. Simul. Eng. Sci. (2015)
92. D. Ryckelynck, T. Goessel, F. Nguyen, Mechanical dissimilarity of defects in welded joints via Grassmann manifold and machine learning. Comptes Rendus. Mécanique **348**(10–11), 911–935 (2020)
93. D. Ryckelynck, L. Hermanns, F. Chinesta, E. Alarcón, An efficient 'a priori' model reduction for boundary element models. Eng. Anal. Bound. Elem. **29**(8), 796–801 (2005)
94. D. Ryckelynck, K. Lampoh, S. Quilicy, Hyper-reduced predictions for lifetime assessment of elasto-plastic structures. Meccanica **51**(2), 309–317 (2016)
95. D. Ryckelynck, D. Missoum Benziane, Multi-level a priori hyper-reduction of mechanical models involving internal variables. Comput. Methods Appl. Mech. Eng. **199**(17), 1134–1142 (2010)

96. D. Ryckelynck, D. Missoum Benziane, Hyper-reduction framework for model calibration in plasticity-induced fatigue. Adv. Model. Simul. Eng. Sci. **3**, (2016)

97. D. Ryckelynck, D. Missoum Benziane, S. Cartel, J. Besson, A robust adaptive model reduction method for damage simulations. Comput. Mater. Sci. **50**(5), 1597–1605 (2011)

98. D. Ryckelynck, F. Vincent, S. Cantournet, Multidimensional a priori hyper-reduction of mechanical models involving internal variables. Comput. Methods Appl. Mech. Eng. **225–228**, 28–43 (2012)

99. B. Sarbandi, S. Cartel, J. Besson et al., Truncated integration for simultaneous simulation of sintering using a separated representation. Arch. Comput. Methods Eng. **17**, 455–463

100. M. Yano, A.T. Patera, An LP empirical quadrature procedure for reduced basis treatment of parametrized nonlinear PDEs. Comput. Methods Appl. Mech. Eng. **344**, 1104–1123 (2019)

101. K. Ye, L.-H. Lim, Schubert varieties and distances between subspaces of different dimensions. SIAM J. Matrix Anal. Appl. **37**(3), 1176–1197 (2016)

102. R. Zimmermann, B. Peherstorfer, K. Willcox, Geometric subspace updates with applications to online adaptive nonlinear model reduction. SIAM J. Matrix Anal. Appl. **39**, 11 (2017)

Open Access This chapter is licensed under the terms of the Creative Commons Attribution 4.0 International License (http://creativecommons.org/licenses/by/4.0/), which permits use, sharing, adaptation, distribution and reproduction in any medium or format, as long as you give appropriate credit to the original author(s) and the source, provide a link to the Creative Commons license and indicate if changes were made.

The images or other third party material in this chapter are included in the chapter's Creative Commons license, unless indicated otherwise in a credit line to the material. If material is not included in the chapter's Creative Commons license and your intended use is not permitted by statutory regulation or exceeds the permitted use, you will need to obtain permission directly from the copyright holder.

Chapter 3
Error Estimation

3.1 Confidence and Trust in Model-Based Engineering Assisted by AI

Consider first data-based machine learning techniques. They rely on large sets of examples provided during the training stage and do not learn with equations. Dealing with a situation that do not belong to the training set variability, namely an out-of-distribution sample, can be very challenging for these techniques. Trusting them could imply being able to guarantee that the training set covers the operational domain of the system to be trained. Besides, data-based AI can lack in robustness: examples have been given of adversarial attacks in which a classifier was tricked to infer a wrong class only by changing a very small percentage of the pixels of the input image. These models often also lack explainability: it is hard to understand what is exactly learned, what phenomenon occurs through the layers of a neural network. In some cases, information on the background of a picture is used by the network in the prediction of the class of an object, or bias present in the training data will be learned by the AI model, like gender bias in recruitment processes. Addressing these limitations is the subject of the Program Confiance.ai,[1] regrouping French academic as well as industrial partners from defense, transports, manufacturing and energy sectors.

Model-based engineering enjoys better explainability–since the reference-model equations are known and used, and robustness–when the reference-model is well-posed. Concerning trust, in our projection-based reduced-order modeling context, the prediction is in general deterministic, and strict error bounds can be derived in particular cases. This is a main difference with AI-based models, which use notions like confidence intervals and predictive uncertainties, expressed as probability results. In

---

[1] https://www.confiance.ai/.

© The Author(s) 2024

D. Ryckelynck et al., *Manifold Learning*, SpringerBriefs in Computer Science,
https://doi.org/10.1007/978-3-031-52764-7_3

the remainder of this chapter, we present error bounds and indicators in projection-based reduced-order modeling, depending on the complexity of the underlying equations.

3.2   In Linear Elasticity and for Linear Problems

Parts of this section has been inspired from the authors previous works [7] and [9].

We suppose that the problem of interest has the following discrete variational form, depending on a parameter μ in a parameter set $\mathcal{P}$: for a finite-dimensional space $\mathcal{V}$ of dimension N (with $N \gg 1$ resulting, e.g., from discretization), find $u_\mu \in \mathcal{V}$ such that

$$E_\mu : a_\mu(u_\mu, v) = b(v), \qquad \forall v \in \mathcal{V}, \tag{3.1}$$

where a_μ is an inf-sup stable bounded sesquilinear form on $\mathcal{V} \times \mathcal{V}$ and b is a continuous linear form on $\mathcal{V}$. We define the Riesz isomorphism J from $\mathcal{V}'$ to $\mathcal{V}$ such that for all $l \in \mathcal{V}'$ and all $u \in \mathcal{V}$, $(Jl, u)_\mathcal{V} = l(u)$, where $(\cdot, \cdot)_\mathcal{V}$ denotes the inner product of $\mathcal{V}$ with associated norm $\| \cdot \|_\mathcal{V}$ and $\mathcal{V}'$ the dual of $\mathcal{V}$. We denote $\beta_\mu := \inf_{u \in \mathcal{V}} \sup_{v \in \mathcal{V}} \frac{|a_\mu(u, v)|}{\|u\|_\mathcal{V} \|v\|_\mathcal{V}} > 0$ the inf-sup constant of a_μ and $\tilde{\beta}_\mu$ a computable positive lower bound of β_μ. For simplicity, we consider that the linear form b is independent of the parameter μ. The extension to μ-dependent b is straightforward.

Applying the Galerkin method on the linear problem (3.1), using a ROB $(\psi_i)_{1 \le i \le n} \in \mathbb{R}^{n \times N}$ as done in Sect. 2.3.2 leads to finding $\hat{u}_\mu \in \mathcal{V}_n$ such that

$$\hat{E}_\mu : a_\mu(\hat{u}_\mu, u_j) = b(u_j), \qquad \forall j \in \{1, \ldots, n\}. \tag{3.2}$$

The approximate solution on the ROB is written as (2.16):

$$\hat{u}_\mu = \sum_{i=1}^{n} \gamma_i^\mu \psi_i. \tag{3.3}$$

We assume that the sesquilinear form a_μ depends on μ in an affine way, namely there exist d functions $\alpha_k(\mu) : \mathcal{P} \to \mathbb{C}$ and d μ-independent sesquilinear forms a_k bounded on $\mathcal{V} \times \mathcal{V}$ such that

$$a_\mu(u, v) = \sum_{k=1}^{d} \alpha_k^\mu a_k(u, v), \qquad \forall u, v \in \mathcal{V}, \tag{3.4}$$

which enables the ROM to be online-efficient.

Under the current assumptions, the following error bound holds (see [16, Sect. 4.3.2]): for all $\mu \in \mathcal{P}$,

$$\|u_\mu - \hat{u}_\mu\|_\mathcal{V} \le \mathcal{E}_1(\mu) := \tilde{\beta}_\mu^{-1}\|G_\mu\hat{u}_\mu\|_\mathcal{V},$$

$$= \tilde{\beta}_\mu^{-1}\left\|G_{00} + \sum_{i=1}^{\hat{N}}\sum_{k=1}^{d}\alpha_k^\mu\gamma_i^\mu G_k u_i\right\|_\mathcal{V}, \tag{3.5}$$

with G_μ the linear map from $\mathcal{V}$ to $\mathcal{V}$ such that $\mathcal{V} \ni u \mapsto G_\mu u := J\left(a_\mu(u, \cdot) - b\right) \in \mathcal{V}$; G_μ inheriting the affine dependence of a_μ on μ since, for all $u \in \mathcal{V}$,

$$G_\mu u = -Jb + \sum_{k=1}^{d}\alpha_k^\mu Ja_k(u, \cdot) = G_{00} + \sum_{k=1}^{d}\alpha_k^\mu G_k u, \tag{3.6}$$

where $G_{00} := -Jb \in \mathcal{V}$ and $G_k u := Ja_k(u, \cdot) \in \mathcal{V}$ for all $k \in \{1, \dots, d\}$.

The error bound (3.5) can rewritten in an equivalent way as

$$\mathcal{E}_2(\mu) = \tilde{\beta}_\mu^{-1}\Bigg((G_{00}, G_{00})_\mathcal{V} + 2\mathrm{Re}\sum_{i=1}^{\hat{N}}\sum_{k=1}^{d}\gamma_i^\mu\alpha_k^\mu(G_k u_i, G_{00})_\mathcal{V}$$

$$+ \sum_{i,j=1}^{\hat{N}}\sum_{k,l=1}^{d}\gamma_i^\mu\alpha_k^\mu\gamma_j^{*\mu}\alpha_l^{*\mu}(G_k u_i, G_l u_j)_\mathcal{V}\Bigg)^{\frac{1}{2}}, \tag{3.7}$$

$$= \tilde{\beta}_\mu^{-1}\left(\delta^2 + 2\mathrm{Re}(s^t\hat{x}_\mu) + \hat{x}_\mu^{*t}S\hat{x}_\mu\right)^{\frac{1}{2}},$$

where $\delta := \|G_{00}\|_\mathcal{V}$, s and $\hat{x}_\mu$ are vectors in $\mathbb{C}^{dn}$ with components $s_I := (G_k u_i, G_{00})_\mathcal{V}$ and $(\hat{x}_\mu)_I := \alpha_k^\mu\gamma_i^\mu$, and S is a matrix in $\mathbb{C}^{dn,dn}$ with coefficients $S_{I,J} := (G_k u_i, G_l u_j)_\mathcal{V}$ (with I and J re-indexing respectively (k, i) and (l, j), for all $1 \le k, l \le d$ and all $1 \le i, j \le n$). The t superscript denotes the transposition. The vector s and the matrix S depend on the ROB $(\psi_i)_{1\le i\le n}$ but are independent of μ, hence can be precomputed; while the vector $\hat{x}_\mu$ depends on the reduced basis approximation $\hat{u}_\mu$ via the coefficients γ_i^μ. A lower bound $\tilde{\beta}_\mu$ of the stability constant of a_μ is also computed in complexity independent of N (which is possible, for example, by the Successive Constraint Method, see [10, 13]).

We would like to stress that $\mathcal{E}_1(\mu) = \mathcal{E}_2(\mu)$ (in infinite precision arithmetic): the indices 1 and 2 are used to denote two different ways to compute the same quantity. In particular, $\mathcal{E}_1(\mu)$ is not online efficient, while $\mathcal{E}_2(\mu)$ is.

3.3 In Nonlinear Mechanics of Materials

The remaining of this section is inspired from the authors work [8].

We look here for an efficient error indicator in the context of general nonlinearities and nonparametrized variabilities in nonlinear structural mechanics. The generality of the assumptions and the complexity of the model lead us to search for quantities correlated to the error made by the ROM with respect to the HFM, instead of rigorous error bounds as considered in the previous section.

The problem of interest is the same as described in Sect. 2.2, and we suppose that we have constructed a ROM following the methods described in Sects. 2.3 and 2.3.5, namely used POD for data compression and ECM for operator compression.

The quantity of interest is the accumulated plastic strain over the complete structure. Since this is a dual quantity, the ROM do not provide directly a prediction over the structure, but only at the reduced quadrature points selected by ECM, see Sect. 2.3.5. The Gappy-POD can be used for to recover the dual quantity of interest over the rest of the structure, see [8, Algorithms 3 and 4] for a presentation on the present context, and [11] for in seminal paper on Gappy-POD.

A quantification for the prediction relative error of the accumulated plastic strain is defined as

$$
E_\mu^p := \begin{cases} \dfrac{\|p_\mu - \tilde{p}_\mu\|_{L^2(\Omega)}}{\|p_\mu\|_{L^2(\Omega)}} & \text{if} \|p_\mu\|_{L^2(\Omega)} \neq 0, \\[2ex] \dfrac{\|p_\mu - \tilde{p}_\mu\|_{L^2(\Omega)}}{\max_{\mu \in \mathcal{P}_{\text{off.}}} \|p_\mu\|_{L^2(\Omega)}} & \text{otherwise,} \end{cases}
\tag{3.8}
$$

where p_μ and $\tilde{p}_\mu$ are respectively the high-fidelity and reduced predictions for the accumulated plasticity field at the variability μ, and $\mathcal{P}_{\text{off.}}$ is the set of variabilities encountered during the offline stage. We underline the fact that $\tilde{p}_\mu$ is the reduced prediction over the complete structure, hence after applying the Gappy-POD reconstruction.

Define the ROM-Gappy-POD residual as

$$
\mathcal{E}_\mu^p := \begin{cases} \dfrac{\|\tilde{\boldsymbol{p}}_\mu - \hat{\boldsymbol{p}}_\mu\|_2}{\|\hat{\boldsymbol{p}}_\mu\|_2} & \text{if} \|\hat{\boldsymbol{p}}_\mu\|_2 \neq 0, \\[2ex] \dfrac{\|\tilde{\boldsymbol{p}}_\mu - \hat{\boldsymbol{p}}_\mu\|_2}{\max_{\mu \in \mathcal{P}_{\text{off.}}} \|\hat{\boldsymbol{p}}_\mu\|_2} & \text{otherwise,} \end{cases}
\tag{3.9}
$$

where $\tilde{\boldsymbol{p}}_\mu$ is the reduced prediction (after applying the Gappy-POD) taken at the reduced quadrature points ($\tilde{p}_{\mu,k} = \tilde{p}_\mu(\hat{x}_k)$, $1 \leq k \leq m^p$), $\hat{\boldsymbol{p}}_\mu$ is the vector of the accumulated plastic strain as computed by the constitutive law solver at the reduced quadrature points during the online stage, and $\| \cdot \|_2$ denotes the Euclidean norm. Notice that in the general case, $\tilde{\boldsymbol{p}}_\mu \neq \hat{\boldsymbol{p}}_\mu$: this discrepancy is at the base of our proposed error indicator.

Notice that the relative error E_μ^p involves fields and L^2-norms whereas the ROM-Gappy-POD residual $\mathcal{E}_\mu^p$ involves vectors of dual quantities in the set of reduced integration points and Euclidean norms. In (3.9), $\|\tilde{\boldsymbol{p}}_\mu - \hat{\boldsymbol{p}}_\mu\|_2$ is the error between the online evaluation of the accumulated plastic strain by the behavior law solver: $\hat{\boldsymbol{p}}_\mu$, and the reconstructed prediction at the reduced integration points $\hat{x}_k$: $\tilde{\boldsymbol{p}}_\mu$, $1 \leq k \leq m^p$. It is explained in [8, Sect. 4.1] that $\|\tilde{\boldsymbol{p}}_\mu - \hat{\boldsymbol{p}}_\mu\|_2$ is also the residual of the least-square optimization involved in the online stage of the Gappy-POD.

Let $B \in \mathbb{R}^{m^p \times n^p}$ such that $B_{k,i} = \psi_i^p(\hat{x}_k)$, $1 \leq k \leq m^p$, $1 \leq i \leq n^p$, $K := \{p_\mu,$ for all possible variabilities $\mu\}$ and $d(K, W)_{L^2(\Omega)} := \sup_{v \in K} \inf_{w \in W} \|v - w\|_{L^2(\Omega)}$, with

W a finite-dimensional subspace of $L^2(\Omega)$. The following propositions and corollary are proven in [8, Sect. 4.1].

Proposition 3.1 *There exist two positive constants C_1 and C_2 independent of μ (but dependent on n^p) such that*

$$\left\| p_\mu - \tilde{p}_\mu \right\|_{L^2(\Omega)}^2 \leq C_1 \| Bz_\mu - \hat{p}_\mu \|_2^2 + C_1 \| p_\mu - \hat{p}_\mu \|_2^2 + C_2 d(K, \text{Span}\{\psi_i^p\}_{1 \leq i \leq n^p})_{L^2(\Omega)}^2. \tag{3.10}$$

Proposition 3.2 *There exist two positive constants K_1 and K_2 independent of μ such that*

$$\left\| \tilde{p}_\mu - \hat{p}_\mu \right\|_2^2 \leq K_1 \left\| p_\mu - \tilde{p}_\mu \right\|_{L^2(\Omega)}^2 + K_2 \| p_\mu - \hat{p}_\mu \|_2^2. \tag{3.11}$$

Corollary 3.3.1 *Suppose that the reduced solution is exact up to the considered time step at the online variability μ: $p_\mu = \tilde{p}_\mu$ in $L^2(\Omega)$. In particular, the behavior law solver has been evaluated with the exact strain tensor and state variables at the integration points x_k, leading to $\hat{p}_\mu(\hat{x}_k) = p_\mu(\hat{x}_k)$, $1 \leq k \leq m^d$. From Proposition 3.2, $\| \tilde{p}_\mu - \hat{p}_\mu \|_2 = 0$, and $\mathcal{E}_\mu^p = 0$.*

We observe that in practice, the evaluations of the ROM-Gappy-POD residual $\mathcal{E}_\mu^p$ (3.9) and the error E_μ^p (3.8) are very correlated in our numerical simulations. The idea is to exploit this correlation by training a Gaussian process regressor for the function $\mathcal{E}_\mu^p \mapsto E_\mu^p$. At the end of the offline stage, we propose to compute reduced predictions at variability values $\{\mu_i\}_{1 \leq i \leq N_c}$ encountered during the data generation step, and the corresponding couples $\left(E_{\mu_i}^p, \mathcal{E}_{\mu_i}^p \right)$, $1 \leq i \leq N_c$. A Gaussian process regressor is trained on these values and we define an approximation function

$$\mathcal{E}_\mu^p \mapsto \text{Gpr}^p(\mathcal{E}_\mu^p), \tag{3.12}$$

for the error E_μ^p at variability μ as the mean plus 3 times the standard deviation of the predictive distribution at the query point $\mathcal{E}_\mu^p$: this is our proposed error indicator. If the dispersion around the learning data is small for certain values $\mathcal{E}_\mu^p$, then adding 3 times the standard deviation will not change very much the prediction, whereas for values with large dispersion of the learning data, this correction aims to provide an error indicator larger than the error. We use the GaussianProcessRegressor python class from scikit-learn [17]. Notice that although some operations in computational complexity dependent on N are carried-out, we are still in the offline stage, and they are much faster than the resolutions of the large size systems of nonlinear equations (2.8). If the offline stage is correctly carried-out and since $\mathcal{E}_\mu^p$ is highly correlated with the error, only small values for $\mathcal{E}_\mu^p$ are expected to be computed. Hence, in order to train the Gaussian process regressor correctly for larger values of the error, the reduced Newton algorithm (2.13) is solved with a large tolerance $\epsilon_{\text{Newton}}^{\text{ROM}} = 0.1$. We call these

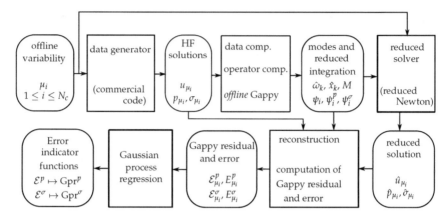

Fig. 3.1 Workflow for the offline stage with error indicator calibration [8]

operations "calibration of the error indication", see Algorithm 3 for a description and Fig. 3.1 for a presentation of the workflow featuring this calibration step.

Algorithm 3: Calibration of the error indicator.

    Input   : Outputs of the data generation, data compression and operator compression steps of Section 2.3.3

    Output: Approximation function $\mathcal{E}_\mu^p \mapsto \mathrm{Gpr}^p(\mathcal{E}_\mu^p)$ of the error

1 Set $\mathcal{Z} = \emptyset$, $k' = 0$, $\hat{\omega} = 0$ and $r_0 = b$;   // initialization

2 **for** $i = 1 \dots, N_c$ **do**

3 | Construct and solve the reduced problem (2.13) with $\epsilon_{\mathrm{Newton}}^{\mathrm{ROM}} = 0.1$

4 | Compute the reconstructed plasticity $\tilde{p}_{\mu_i}$ using Gappy-POD and $\mathcal{E}_{\mu_i}^p$

5 | Compute the error $E_{\mu_i}^p$ using (3.8)

6 | $\mathcal{X} \leftarrow \mathcal{X} \cup \left(\mathcal{E}_{\mu_i}^p, E_{\mu_i}^p\right)$

7 Construct an approximation function $\mathcal{E}_\mu^p \mapsto \mathrm{Gpr}^p(\mathcal{E}_\mu^p)$ of the error E_μ^p using a Gaussian process regression and the data from $\mathcal{X}$

We recall that in model order reduction, the original hypothesis is the existence of a low-dimensional vector space where an acceptable approximation of the high-fidelity solution lies. The hypothesis is formalized under a rate of decrease for the Kolmogorov n-width with respect to the dimension of this vector space. The same hypothesis is made when using the Gappy-POD to reconstruct the dual quantities, which are expressed as a linear combination of constructed modes. For both the primal and dual quantities, the modes are computed by searching some low-rank structure of the high-fidelity data. The coefficients of the linear combination for reconstructing the primal quantities are given by the solution of the reduced Newton algorithm (2.13). After convergence, the residual is small, even in cases where the reduced order model exhibits large errors with respect to the high-fidelity reference: this residual gives no information on the distance between the reduced solution and the high-fidelity finite element space.

However, in the online phase of the ROM-Gappy-POD reconstruction (see [8, Algorithm 4]), the coefficients $\hat{p}_{\mu,k}$ (the accumulated plastic strain computed by the constitutive law solver during the online stage) contain information from the high-fidelity behavior law solver. Moreover, an overdetermined least-square is solved, which can provide a nonzero residual that implicitly contains this information from the high-fidelity behavior law solver: namely the distance between the prediction from the behavior law and the vector space spanned by the Gappy-POD modes (restricted to the reduced integration points): this is the term $\|Bz_\mu - \hat{p}_\mu\|_2$ in (3.10). Hence, the ability of the online variability to be expressed on the Gappy-POD modes is monitored through the behavior law solver on the reduced integration points. When the ROM is solved for an online variability not included in the offline variabilities, then the new physical solution cannot be correctly interpolated using the POD and Gappy-POD modes: hence, the ROM-Gappy-residual becomes large.

From Proposition 3.2, if $\|Bz_\mu - \hat{p}_\mu\|_2 = \|\tilde{p}_\mu - \hat{p}_\mu\|_2$ is large, then the global error $\|p_\mu - \tilde{p}_\mu\|_{L^2(\Omega)}$ and/or the error at the reduced integration points $\hat{x}_k$ is large, which makes $\|Bz_\mu - \hat{p}_\mu\|_2$ a good candidate for a enrichment criterion for the ROM. A limitation of the error indicator can occur if the online variability activates strong nonlinearities on areas containing no point from the reduced integration scheme, namely through the term $C_2 d(K, \text{Span}\{\psi_i^p\}_{1 \le i \le n^p})_{L^2(\Omega)}^2$ in (3.10).

We recall that the error indicator (3.12) is a regression of the function $\mathcal{E}_\mu^p \mapsto E_\mu^p$. In the online phase, we only need to evaluate $\mathcal{E}_\mu^p$ and do not require any estimation for the other terms and constants appearing in Propositions 3.1 and 3.2.

Equipped with an efficient error indicator, we are now able to assess the quality of the approximation made by the reduced order model in the online phase. If the error indicator is too large, an enrichment step occurs: the high-fidelity model is used to compute a new high-fidelity snapshot, which is used to update the POD and Gappy-POD basis, as well as the reduced integration schemes. Notice that for the enrichment steps to be computed, the displacement field and all the state variables of the previous time step need to be reconstructed on the complete mesh Ω to provide the high-fidelity solver with the correct material state. The workflow for the *online* stage with enrichment is presented in Fig. 3.2.

We refer to [8] for more details on this subject, and detailed numerical applications in nonlinear structural mechanics for this error indicator and its ability to enrich a ROM in the online stage.

Notice that another (noncertified) indicator in nonlinear solid mechanics with internal variables has been proposed in [1], aiming to approximate the dual norm of residuals in the same fashion as in the linear case described in Sect. 3.2. For such nonlinear case, rigorous error bounds are not obtained: a gappy-POD-based approximate representation of the stress tensor is used, and the inf-sup constant evaluation has been replaced by a normalization of the residual using the norm of the Riesz elements for the external loading.

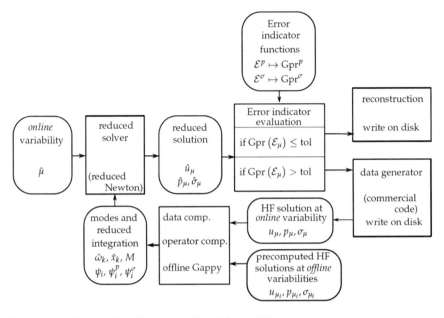

Fig. 3.2 Workflow for the online stage with enrichment [8]

3.4 In Computational Fluid Dynamics

In the following section, we present a priori error estimates due to the POD-Galerkin approximation applied to fluid dynamics equations in particular and parabolic nonlinear PDEs in general. It is a theoretical result on the convergence of the POD-Galerkin reduced order model towards the high-fidelity semi-discretized equations in the sense of the spatial variable. The solution of these semi-discretized equations is denoted $\tilde{u}^h$ over a time interval $[0, T]$ such that $h = \frac{1}{M}$ and M is the cardinal of a Hilbert basis that is capable of generating the high-fidelity semi-discretized solution at some specified M time instants. This orthonormal basis can be obtained in an a posteriori fashion thanks to the snapshots POD method. It is denoted by $(\psi_i)_{i=1,\ldots,M}$.

The following convergence result includes furthermore a discussion on the stability of the Galerkin projection technique applied to parabolic PDEs. This result has been developed and published in the following papers: [2–5].

In the literature, we can find many works around the problem of defining convergent and a priori upper bounds for the POD-Galerkin reduced models for parabolic equations. It is a subject of great interest, if we can quantify efficiently the error of a reduced solution $\widehat{u}$ obtained with an approximation technique of the corresponding high fidelity semi-discretized solution $\tilde{u}^h$. This problem can be seen as a theoretical confidence interval around a training point of an approximation model (details on some convergence results of the literature should be added). Let us denote by Ω the open and bounded domain of the spatial variable, such that $\Omega \in \mathbb{R}^d$, where

$d = 1$ *or* 2. Let us consider the parabolic PDEs for which the weak formulation in the space of the solution u^h spanned by the POD basis of cardinal M, is described as follows:

- b is a trilinear form defined over $[H^1(\Omega)]^d \times [H^1(\Omega)]^d \times [H^1(\Omega)]^d$,
- a dissipating term defined as a bi-linear and coercive form a over $[H^1(\Omega)]^d \times [H^1(\Omega)]^d$,
- a linear form F_t defined over $[L^2(\Omega)]^d$,
- β is the coercivity constant of the bilinear form a,
- C_a and C_b are respectively the norms of a and b in the space $[L^2(\Omega)]^d$,
- $K = \left\| u^h \right\|_{L^\infty(0,T;[L^2(\Omega)]^d)}$.

The following result is proved:

Theorem 3.1 *If $n << M$ is the dimension of the truncated POD-Galerkin reduced solution $\widehat{u}$ that represent the training point (solution) u^h, then we can derive the following a priori upper bound of the $[L^2(\Omega)]^d-$ error between $\widehat{u}(t)$ and $\widetilde{u}^h(t)$, $\forall t \in [0, T]$:*

$$\left\| (\widetilde{u}^h - \widehat{u})(t) \right\|^2_{[L^2(\Omega)]^d} \le f_1(N). \tag{3.13}$$

Where $f_1(n)$ is the remainder of the sum of a convergent series; $f_1(n)$ converges to 0 when N converges to M. More precisely, $f_1(n)$ is a function of the remainder of the sum of the POD eigenvalues $(\lambda_k)_{k=1,...,M}$ obtained from the Snapshots POD applied to M temporal snapshots of u^h: it is expressed into two different fashions: Either

$$f_1(n) = \left(1 + 2C_bK + \frac{C_a^2}{\epsilon} \right) T \sum_{k=n+1}^{M} \lambda_k, \tag{3.14}$$

if $(\epsilon - 2\beta + 6C_bK) \le 0$, for a strictly positive real number ϵ, or

$$f_1(n) = \left(T + \left(2C_bK + \frac{C_a^2}{\epsilon} \right) T \exp(T[2C_aK + K^2(1 + C_a^2)]) \right) \sum_{k=n+1}^{M} \lambda_k, \tag{3.15}$$

if $(\epsilon - 2\beta + 6C_bK) > 0$, for all strictly positive real numbers ϵ.

The mathematical proof of Theorem 3.1 is based on the properties of the forms a and b, the application of the Young inequality and the Gronwall lemma. For the details of the proof, refer to [4].

Remark 3.1 The result of Theorem 3.1 is applicable in particular for the $1D$ Burgers equation and the $2D$ unsteady and incompressible Navier-Stokes equations, by remarking the following:

- When $d = 1$, $H^1(\Omega) \subset L^\infty(\Omega)$: there exists $C_\infty^1 \in \mathbb{R}^{+*}$ such that $\forall v \in H^1(\Omega)$, $\|v\|_{L^\infty(\Omega)} \le C_\infty^1 \|\nabla v\|_{L^2(\Omega)}$.

- When $d = 2$ or $d = 3$, $H^1(\Omega) \subset L^4(\Omega)$: there exists $C_4^1 \in \mathbb{R}^{+*}$ such that $\forall\ v \in H^1(\Omega)$, $\|v\|_{L^4(\Omega)} \leq C_4^1 \|\nabla v\|_{L^2(\Omega)}$.
- When $d = 2$: there exists $C \in \mathbb{R}^{+*}$ such that:

$$\|v\|_{L^4(\Omega)} \leq C \|v\|_{L^2(\Omega)}^{1/2} \|\nabla v\|_{L^2(\Omega)}^{1/2} .$$

- If we denote by S the square matrix of dimension M defined by: $S_{i,j} = \langle \psi_i, \psi_j \rangle_{[H^1(\Omega)]^d}$, then $\forall\ v$ in $\mathcal{V}^h$ the solutions space spanned by the complete POD basis ψ we have the following inequality: $\|v\|_{[H^1(\Omega)]^d} \leq \sqrt{\|S\|} \|v\|_{[L^2(\Omega)]^d}$. Where $\|S\|$ is the spectral norm of the matrix S. For details, refer to [14].

Remark 3.2 A result of stability with respect to time of the POD-Galerkin reduced model for nonlinear and dissipated parabolic PDEs can be derived from result 3.1. If μ denotes only the viscosity constant in this particular case (without any loss of generality), then the POD-Galerkin reduced model is stable with respect to time when μ satisfies the following inequality:

$$\mu \geq \frac{6C_b K + \epsilon}{2\beta},$$

where ϵ is a strictly positive real number.

Remark 3.3 More particularly, in the case of linear and dissipated parabolic PDEs, the stability condition of the POD-Galerkin reduced model becomes equivalent to $\mu \geq \frac{\epsilon}{2c_p}$. The error of the POD-Galerkin reduced model with respect to the high fidelity training point can be estimated exactly as in the Céa lemma applied for elliptic and sesquilinear PDEs: if $\mu \geq 1$ then,

$$\left\| (u^h - \widehat{u})(t) \right\|_{[L^2(\Omega)]^d}^2 \leq \left(1 + \frac{C_a^2}{2\beta} \right) T \sum_{n=N+1}^{M} \lambda_n.$$

Based on the same methodology, we propose in what follows an a priori error estimate and a convergence result for POD-Galerkin reduced model parameter wise. In other words, we show that when the parameters change with respect to the training ones, a confidence interval is obtained around the new test solution. The width of this confidence interval converges to the truncation error of the ROM at the training parameters. This parametric convergence result is formulated as follows:

Theorem 3.2 *Let us denote by $u_{\mu_0}^h$ a training solution associated with a training parameter μ_0. So we suppose we have only one training point for the POD-Galerkin reduced model, without any loss of generality. If ψ^{μ_0} denotes the Hilbert basis obtained from the POD applied to the high fidelity training solution and $\widehat{u}_{\mu,\mu_0}$ denotes the truncated POD-Galerkin reduced solution that approaches the test point (solution) u_μ^h in the reduced POD space of dimension n spanned by ψ^{μ_0}, then:*

$$\left\| \left(u_\mu^h - \widehat{u}_{\mu,\mu_0} \right)(t) \right\|^2 \leq f_1^{\mu_0}(n) \left(1 + \| \mu - \mu_0 \|^{2\alpha} \right) + f_2^{\mu_0}(n) \, \| \mu - \mu_0 \|^{\alpha}, \quad (3.16)$$

where $f_2^{\mu_0}(n)$ is the remainder of the sum of a convergent series; $f_2^{\mu_0}(n)$ converges to 0 when n converges to M. More precisely $f_2^{\mu_0}(n)$ is a function of the remainder of the sum of the orthogonal projection coefficients of the characteristic function 1_{Ω_x} such that Ω_x tends to Ω when x tends to $\partial\Omega$: it is expressed as follows:

$$f_2^{\mu_0}(n) = \left(\left\| (u_\mu^h - u_{\mu_0}^h)(0) \right\|^{-4} + \frac{B}{A}(\exp(-2At) - 1) \right)^{-1/2} C_a^2 \left\| \nabla u_{\mu_0}^h \right\|^2 \sum_{k=n+1}^{M} \langle 1_{\Omega_x}, \psi_k^{\mu_0} \rangle^2,$$

$$(3.17)$$

where A and B are strictly positive constants of which the detailed expressions are given in [2].

The proof of the above theorem is published in [2]. It is based on the properties of the two forms a and b, the application of the Young inequality and the resolution of a nonlinear ordinary differential inequality of Ricatti type.

3.5   A Note on Accuracy of a Posteriori Error Bounds and Round-Off Errors

In this section, we explain why the online-efficient error bound (3.7) may be sensitive to round-off errors.

In computers, real numbers are represented by a finite number of bits, called floating-point representation. Current architectures are optimized for the IEEE 754 double-precision binary floating-point format. Let x and y be real numbers. When computing the operation $x + y$, the result returned by the computer can be different from its theoretical value. Whenever the difference is substantial, a loss of significance occurs. A well-known case of loss of significance is when x and y are almost opposite numbers. Suppose that $x = -y$. We denote by maxfl$(x + y)$ the result that the computer returns when the maximal accumulation of round-off errors occurs when computing the summation. There holds

$$|\text{maxfl}(x + y)| \approx 2\epsilon |x|, \quad (3.18)$$

where ϵ is called the machine precision. In double precision, $\epsilon = 5 \times 10^{-16}$ (see [12, Sect. 1.2]).

When implementing an algorithm, one should ensure that each step is free of such a loss of significance. In some cases, simply changing the order of the operations can prevent these situations. As an illustration, consider $x = 1$, $y = 1 + 10^{-7}$, and the operation $x^2 - 2xy + y^2$. This is a sum of terms where the first intermediate result in the sum is 14 orders larger than the result. Therefore, a loss of significance is expected. The relative error of this computation is about 8×10^{-4}. Computing $(x - y)^2$, which is the factorization of the considered operation, leads to a relative error of about 10^{-9}. Thus, the terms of the sum are only 7 orders larger than the results, leading to a less catastrophic loss of significance. In this specific case, the remedy consists in carrying out the sum before the multiplication. In our projection-based ROM context, the evaluation of the formula $\mathcal{E}_2$ suffers from such a loss of significance, as we now explain.

We investigate the influence of round-off errors when computing the error bounds (3.5) and (3.7) for respectively $\mathcal{E}_1(\mu)$ and $\mathcal{E}_2(\mu)$. As observed in the previous paragraph, the computation of a polynomial using a factorized form is more accurate than using the developed form, in particular at points close to its roots. Here, $\left(\tilde{\beta}_\mu\mathcal{E}_2(\mu)\right)^2$ is a multivariate polynomial of degree 2 in $\hat{x}_\mu$ computed in a developed form, whereas the scalar product $(G_\mu u_\mu, G_\mu u_\mu)_V$ used in the computation of $\mathcal{E}_1(\mu)$ is not developed. The following holds (see [9, Proposition 2.2.1] for the proof)

Proposition 3.3 *Let $\mu \in \mathcal{P}$ and let* $\mathrm{maxfl}(\tilde{\beta}_\mu\mathcal{E}_k(\mu))$, $k = 1, 2$, *denote the evaluation of $\tilde{\beta}_\mu\mathcal{E}_k(\mu)$ when the maximum accumulation of round-off errors occurs. There holds*

$$
\begin{aligned}
\mathrm{maxfl}(\tilde{\beta}_\mu\mathcal{E}_1(\mu)) &\geq 2\delta\epsilon, \\
\mathrm{maxfl}(\tilde{\beta}_\mu\mathcal{E}_2(\mu)) &\geq 2\delta\sqrt{\epsilon},
\end{aligned}
\tag{3.19}
$$

where $\delta = \|G_{00}\|_V$ and ϵ is the machine precision.

From this proposition, we notice that the online-efficient formula $\mathcal{E}_2(\mu)$ suffers from an important loss of significance.

We present below an error estimator proposed in [9] that enjoys both accuracy and online-efficiency. Let $\sigma := 1 + 2dn + (dn)^2$. For a given $\mu \in \mathcal{P}_{\mathrm{trial}}$ and the resulting $\hat{u}_\mu \in \mathrm{Span}\{\psi_1, ..., \psi_n\}$ solving the reduced problem (3.2), we define $\hat{X}(\mu) \in \mathbb{C}^\sigma$ as the vector with components $(1, \hat{x}_{\mu_I}, \hat{x}^*_{\mu_I}, \hat{x}^*_{\mu_I}\hat{x}_{\mu_J})$, where $\hat{x}_{\mu_I} = \alpha_k^\mu \gamma_i^\mu$ (we recall that γ_i^μ are the coefficients of the reduced solution in the reduced basis, see (3.3), and α_k^μ the coefficients of the affine decomposition of a_μ in (3.4)), with $1 \leq I, J \leq dn$ (with $I = i + n(k-1)$ such that $1 \leq i \leq n, 1 \leq k \leq d$, and with $J = j + n(l-1)$ such that $1 \leq j \leq n, 1 \leq l \leq d$). We can write the right-hand side of (3.7) as a linear form in $\hat{X}(\mu)$ as follows:

$$
\delta^2 + 2\mathrm{Re}(s^t\hat{x}_\mu) + \hat{x}^{*t}_\mu S\hat{x}_\mu = \sum_{p=1}^\sigma t_p \hat{X}_p(\mu),
\tag{3.20}
$$

where t_p is independent of μ (as δ, s, and S are independent of μ) and $\hat{X}_p(\mu)$ is the p-th component of $\hat{X}(\mu)$.

Consider the function of two variables $(p, \mu) \mapsto \hat{X}_p(\mu)$, for all $p \in \{1, ..., \sigma\}$ and all $\mu \in \mathcal{P}$. We look for an approximation of this function in the form

$$
\forall \mu \in \mathcal{P}, \forall p \in \{1, ..., \sigma\}, \ \hat{X}_p(\mu) \approx \sum_{r=1}^{\hat{\sigma}} \lambda_r^{\hat{\sigma}}(\mu)\hat{X}_p(\mu_r),
\tag{3.21}
$$

for a certain parameter $\hat{\sigma} \leq \sigma$. The Empirical Interpolation Method (EIM) provides a numerical procedure to construct this approximation and to choose the interpolation points (see [6, 15]), which leads to the following formula for computing the error bound

$$\mathcal{E}_3(\mu) := \tilde{\beta}_\mu^{-1} \left(\sum_{r=1}^{\hat{\sigma}} \lambda_r^{\hat{\sigma}}(\mu) V_r \right)^{\frac{1}{2}}, \tag{3.22}$$

where $V_r = \left\| G_{\mu_r} \hat{u}_{\mu_r} \right\|_V^2$, and where $\lambda_r^{\hat{\sigma}}(\mu)$ and μ_r are provided by EIM, see [9, Sect. 3.2] for all the details of this derivation. There holds (see [9, Proposition 3.2.1]):

Proposition 3.4 *The computation of the formula $\mathcal{E}_3$ is well defined, and this formula is online-efficient.*

Besides, $\mathcal{E}_3$ involving a linear combination of accurately computed scalar products (see (3.5)), it is not subject to the loss of significance encountered in $\mathcal{E}_2$.

We refer to [9] for more details on the notion of validity of a formula to compute an error bound, additional variants for accurate and efficient error bounds (including one featuring a stabilized EIM), as well as numerical illustrations for a one-dimensional linear diffusion problem and and three-dimensional acoustic scattering problem. The error bound $\mathcal{E}_3$ can be of particular interest in the following situations: (i) when the stability constant of the original problem is very small (this is the case in many practical problems), (ii) when very accurate solutions are needed, (iii) when considering a nonlinear problem (for which, in some cases, no error bound is possible until a very tight tolerance is reached, see [18]).

References

1. E. Agouzal, J-P. Argaud, M. Bergmann, G. Ferté, T. Taddei, A projection-based reduced-order model for parametric quasi-static nonlinear mechanics using an open-source industrial code (2022)
2. N. Akkari, A. Hamdouni, L. Erwan, M. Jazar, On the sensitivity of the pod technique for a parameterized quasi-nonlinear parabolic equation. Adv. Model. Simul. Eng. Sci. **1**, 14, 08 (2014)
3. N. Akkari, A. Hamdouni, M. Jazar, Mathematical and numerical results on the parametric sensitivity of a rom-pod of the burgers equation. Eur. J. Comput. Mech. **23**(1–2), 78–95 (2014)
4. N. Akkari, A. Hamdouni, M. Jazar, Mathematical and numerical results on the sensitivity of the pod approximation relative to the burgers equation. Appl. Math. Comput. **247**, 951–961 (2014)
5. N. Akkari, A. Hamdouni, E. Liberge, M. Jazar, A mathematical and numerical study of the sensitivity of a reduced order model by pod (rom–pod), for a 2d incompressible fluid flow. J. Comput. Appl. Math. **270**, 522–530 (2014), in *Fourth International Conference on Finite Element Methods in Engineering and Sciences (FEMTEC 2013)*
6. M. Barrault, Y. Maday, N.C. Nguyen, A.T. Patera, An empirical interpolation method: application to efficient reduced-basis discretization of partial differential equations. Comptes Rendus Mathematiques **339**(9), 666–672 (2004)

7. F. Casenave, Accurate a posteriori error evaluation in the reduced basis method. Comptes Rendus Mathematique **350**(9–10), 539–542 (2012)
8. F. Casenave, N. Akkari, An error indicator-based adaptive reduced order model for nonlinear structural mechanics - application to high-pressure turbine blades. Math. Comput. Appl. **24**(2), (2019)
9. F. Casenave, A. Ern, T. Lelièvre, Accurate and online-efficient evaluation of the a posteriori error bound in the reduced basis method. ESAIM Math. Model. Numer. Anal. **48**(1), 207–229 (2014)
10. Y. Chen, J.S. Hesthaven, Y. Maday, J. Rodríguez, Improved successive constraint method based a posteriori error estimate for reduced basis approximation of 2d Maxwell's problem. ESAIM Math. Model. Numer. Anal. **43**(6), 1099–1116, 8 (2009)
11. R. Everson, L. Sirovich, Karhunen-Loève procedure for gappy data. J. Opt. Soc. Am. A **12**(8), 1657–1664 (1995)
12. D. Goldberg, What every computer scientist should know about floating point arithmetic. ACM Comput. Surv. **23**(1), 5–48 (1991)
13. D.B.P. Huynh, G. Rozza, S. Sen, A.T. Patera, A successive constraint linear optimization method for lower bounds of parametric coercivity and inf-sup stability constants. Comptes Rendus Mathematique **345**(8), 473–478 (2007)
14. K. Kunisch, S. Volkwein, Galerkin proper orthogonal decomposition methods for parabolic problems. Numerische mathematik **90**(1), 117–148 (2001)
15. Y. Maday, N.-C. Nguyen, A.T. Patera, S.H. Pau, A general multipurpose interpolation procedure: the magic points. Commun. Pure Appl. Anal. **8**(1), 383–404 (2009)
16. A.T. Patera, G. Rozza, Reduced Basis Approximation and A Posteriori Error Estimation for Parametrized Partial Differential Equations. MIT Pappalardo Graduate Monographs in Mechanical Engineering (2007)
17. F. Pedregosa, G. Varoquaux, A. Gramfort, V. Michel, B. Thirion, O. Grisel, M. Blondel, P. Prettenhofer, R. Weiss, V. Dubourg, J. Vanderplas, A. Passos, D. Cournapeau, M. Brucher, M. Perrot, E. Duchesnay, Scikit-learn: machine learning in Python. J. Mach. Learn. Res. **12**, 2825–2830 (2011)
18. M. Yano, A space-time Petrov-Galerkin certified reduced basis method: application to the Boussinesq equations. SIAM J. Sci. Comput. **36**(1), A232–A266 (2014)

Open Access This chapter is licensed under the terms of the Creative Commons Attribution 4.0 International License (http://creativecommons.org/licenses/by/4.0/), which permits use, sharing, adaptation, distribution and reproduction in any medium or format, as long as you give appropriate credit to the original author(s) and the source, provide a link to the Creative Commons license and indicate if changes were made.

The images or other third party material in this chapter are included in the chapter's Creative Commons license, unless indicated otherwise in a credit line to the material. If material is not included in the chapter's Creative Commons license and your intended use is not permitted by statutory regulation or exceeds the permitted use, you will need to obtain permission directly from the copyright holder.

Chapter 4
Resources: Software and Tutorials

4.1 Mordicus: Reduced-Order Methods Designed for Industrial Usage

MORDICUS is the French acronym for "Méthodes d'Ordre RéDuIt Conçues pour des Usages induStriels", translated to "reduced-order methods designed for industrial usage". It is the name of a collaborative project that took place from 2018 to 2023, with the objective of developing a standard for a datamodel and basic computational treatment for reduced-order modeling in the French community.

4.1.1 Mordicus Project and Consortium

"MOR_DICUS" is a project of type FUI (Fonds Unique Interministériel), in which took part a consortium of ten partners of small and large companies, as well as research institutes and universities. More precisely, the partners are Électric-ité de France (EDF), Safran Group, ESI Group, Hexagon, Phimeca, Transvalor, CT Ingénierie, CEMOSIS, Armines and Laboratoire Jacques-Louis Lions (LJLL-UPMC), see Fig. 4.1.

The founding is provided by BPI France, pôle systematic and pôle Alsace énergieVie, see Fig. 4.2.

4.1.2 Mordicus Library

Mordicus consists in a Python library and a C++ application, where a datamodel and common basic algorithms and tools are proposed to numerically carry out surrogate modeling and projection-based ROM.

© The Author(s) 2024
D. Ryckelynck et al., *Manifold Learning*, SpringerBriefs in Computer Science,
https://doi.org/10.1007/978-3-031-52764-7_4

Fig. 4.1 Partners of the MOR_DICUS consortium

Fig. 4.2 Funders of the MOR_DICUS consortium

The sources of the library are available on GitLab.com[1] and the user's license is the GNU LGPLv3. A website containing an installation procedure, a description of the library, some tutorial and a documentation is also available.[2] The rest of the section is inspired from the "Numerical methods"[3] section of this documentation.

The Mordicus library is constructed with the datamodel at its center, and the methods and algorithms act as functions that can modify the state of the datamodel. As detailed in Sect. 2.3, Reduced-Order Model workflows involve an offline (or learning) stage, and an online (or exploitation) stage. The different steps can be vastly adapted and mixed together to produce any complex workflows, as long as the algorithm makes sense. For instance, a regressor can be trained directly on high-dimensional data, and the data compression step would not exist. Another example is the Reduced-Basis method, where the notions of offline and online stages merge together, since HFM and ROM resolutions are alternatively computed when constructing the reduced-order basis.

Mordicus provides a datamodel, standard interfaces and basic algorithms, as illustrated in Fig. 4.3. Based on Mordicus, applicative modules can then be developed and propose new workflows and algorithms. The library genericROM, described in Sect. 4.2, is an applicative module of Mordicus.

Simple algorithms are proposed, and can be used by any applicative module.

- SVD: for computing truncated singular value decomposition of lower triangular matrices.

---

[1] https://gitlab.com/mor_dicus/mordicus.

[2] https://mordicus.readthedocs.io.

[3] https://mordicus.readthedocs.io/en/latest/_methods/index.html.

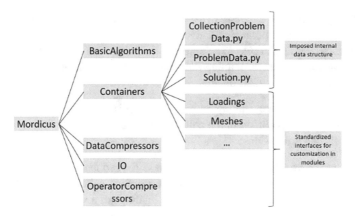

Fig. 4.3 Organization and overview of the datamodel of the Mordicus library

- `ScikitLearnRegressor`: contains examples for computing grid search cross validation and a customizable Gaussian Process Regressor using scikit-learn tools.
- `Interpolation`: contains efficient time Interpolation tool used in the Mordicus datamodel.

More details on the datamodel are provided in the next Section.

4.1.3 Mordicus Datamodel

The main feature of Mordicus is a datamodel adapted to reduced-order modeling. It has been constructed to facilitate collaboration, by proposing three main classes `CollectionProblemData`, `ProblemData` and `Solution`, supposed to be populated and handle in the same fashion by all users (see Fig. 4.3):

- `CollectionProblemData`: The meta structure containing the complete data-model for a reduced-order model.
- `ProblemData`: Containing a model for a physics problem: initial condition, loading, constitutive laws, solutions.
- `Solution`: Containing the size, snapshots and reduced coordinated of solutions.

These classes also contain numerous functions to iterate, modify and handle the data structure. We present a few important ones below:

- `CollectionProblemData` functions:

  - `DefineVariabilityAxes()`: sets the axes of variability, can be strings for nonparametrized variability, or floats,
  - `SnapshotsIterator()`: returns an iterator over snapshots of solutions of a given name,

  – `GetSnapshotsAtTimes()`: returns an array containing all the snapshots of
    a given name, interpolated at a given time,
  – `CompressSolutions()`: compress the snapshots of solutions of a given
    name against the corresponding reduced-order basis, and update to correspond-
    ing solution.reducedCoordinates.

  Notice that the functions acting on solution objects automatically iterate over all
  the problemDatas included in the collectionProblemData.
- `ProblemData` functions:

  – `UncompressSolution()`: uncompress the reducedCoordinates of a solu-
    tion of a given name, and update to corresponding solution.snapshots,
  – `GetLoadingsOfType()`: returns all loadings of a specific type in a list,
  – `GetParameterAtTime()`: returns the parameter value at a specitiy time
    (with time interpolation if needed).

- `Solution` functions:

  – `UncompressSnapshotAtTime()`: uncompress the reducedCoordinates of
    the solution at a given time, and update to corresponding snapshots,
  – `GetTimeSequenceFromSnapshots()`: returns the time sequence from
    the snapshots dictionary,
  – `ConvertReducedCoordinatesReducedOrderBasisAtTime()`:
    converts the reducedSnapshot at a given time from the current reducedOrder-
    Basis to a newReducedOrderBasis using a projectedReducedOrderBasis.

  The data-model has also been thought to be agile and customizable, by allowing
developers to propose other classes, in their applicative module, or variant classes of
the ones contained in the subfolders of Containers (e.g. Loadings, Meshes).

4.2 GenericROM Library

The genericROM software consists in a Python library, acting as an applicative
module of Mordicus, and developed at Safran.

The sources of the library are available on GitLab.com[4] and the user's license
is BSD 3-Clause, a very permissive license that allows the users to use and redis-
tribute the sources, with or without modification. A website containing an installation
process, a description of the library, some tutorial and a documentation are also available.[5]

---

[4] https://gitlab.com/drti/genericrom.

[5] https://genericrom.readthedocs.io.

4.2.1 Main Available Methods

We refer to Sect. 2.3 for an description of the main steps of projection-based reduced-order modeling methods.

The simplest data compression method available in genericROM is the snapshot-POD, as described in Sect. 2.3.3.2. An implementation is available in parallel with distributed memory, by partitioning the domain Ω in subdomains, as well as an incremental version.

The only hyper-reduction method available to date in genericROM is the ECM, see Sect 2.3.5. A variant of th Nonnegative Orthogonal Matching Pursuit Algorithm 2.2 is implemented, where randomly chosen quadrature point can be added at each iteration to prevent the algorithm to fall into local minima.

The physical problems available for reduction by genericROM are

- nonlinear quasistatic structural mechanics, with possibibly very complex constitutive laws, centrifugal and temperature effects, as well as pressure and homogeneous Dirichlet boundary conditions;
- transient thermal problems, with heat capacity and conductivity depending on the solution temperature in a nonlinear fashion, as well as convection heat flux and radiation boundary conditions.

4.2.2 Noninstrusivity and Nonparametrized Variability

In genericROM, projection-based ROMs can be constructed even when the snapshots are generated by commercial software, as long as readers for the meshes and computed snapshots are available (or can be developed). The handling of projection-based ROM workflows without having to modify the assembling routines of the reference HFM code is the reason why the library is coined nonintrusive. To do so, the assembling routine of operators and right-hand sides are handling directly in genericROM, using the finite element engine of BasicTools[6] developed at Safran.

This nonintrusive feature is rare in the literature, and has important advantages in an industrial context: expensive and already computed databases of snapshots can be used to construct a ROM. Besides, the ROM library can be updated without having to undergo complex certification of the evolution of reference HFM codes. Moreover, when these codes are commercial, intrusive implementations of projection-based ROM are not possible, or may require expensive contracts with code editors. An example of a ROM constructed with genericROM for a nonlinear transient thermal problem using snapshots computed by Abaqus is available in [2].

GenericROM can also deal with nonparametrized variability, in the sense that the variability at the origin of the differences in the snapshots computed in the data

---

[6] https://gitlab.com/drti/basic-tools.

generation step is not required to be formalized, or even known. This variability takes usually the form of parameters in the definition of the reference HFM, but can also be unknown when for instance the boundary conditions come from a first numerical simulation. Besides, one may want to compute a reduced model by imposing a variability outside a parametrization used to generate the database of snapshots. The nonparametrized variability feature comes with an efficiency trade-off, since, for a certain variability to be changed in a nonparametrized fashion, the ROM needs to assemble the uncompressed version of the corresponding terms in the weak form, before compressing them in the reduced problem (2.13). GenericROM offers the possibility to precompute all the parametrized variability, in order for the ROM to compute efficiently the corresponding terms in the reduced problem, with the consequence that such terms can only be updated following this parametrization. More details on the efficiency are given in the next section. The next two sections are inspired from the "Numerical methods"[7] and "Tutorials"[8] section of the genericROM documentation.

4.2.3 Precomputations for Efficiency

In ROM workflows, the common measure of efficiency is the speedup, defined as the ratio between the computation duration of the HFM and the one of the ROM. For a given accuracy, the higher the speedup, the better. Various elements can be taken into account, including code optimization of the online stage. The methods enabling algorithmic complexity gains have been presented in Sect. 2.3.3. In this section, we give additional implementation elements and illustrations to give some intuition on the mechanisms at play.

Depending on the considered equations and nature of variability, the goal is to precompute as many quantities as possible in the offline stage, to leave as few operations as possible to the online stage. In the HFM, the assembling of the global tangent operator at each Newton iteration, using the high-fidelity quadrature, reads:

$$\frac{D\mathcal{F}_\mu}{Du}\left(u^k\right)_{ij} = \sum_{g=1}^{N_g} \omega_g \left[\epsilon\left(\varphi_j\right) : \mathcal{K}\left(\epsilon(u_\mu^k), y_\mu\right) : \epsilon\left(\varphi_i\right)\right]\left(x_g\right), \ 1 \leq i, j \leq N. \ (4.1)$$

In the ROMs implemented in genericROM, the assembling of the reduced global tangent operator at each Newton iteration, using the reduced quadrature computed by ECM, reads:

---

[7] https://genericrom.readthedocs.io/en/latest/_methods/index.html.
[8] https://genericrom.readthedocs.io/en/latest/_tutorials/Mechanical/MecaSequential/index.html.

Fig. 4.4 HFM: a large
sparse linear system is
assembled and solved at each
step of the Newton algorithm

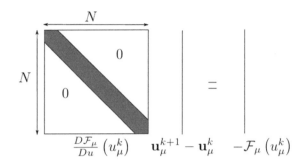

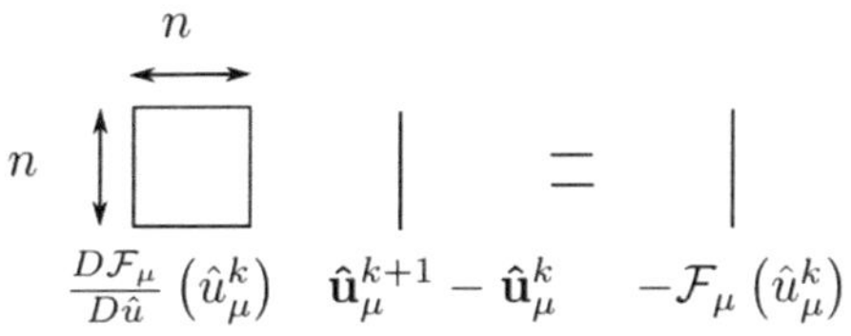

Fig. 4.5 ROM: a small
dense linear system is
assembled and solved at each
step of the reduced Newton
algorithm

$$\frac{D\mathcal{F}_\mu}{D\hat{u}}\left(\hat{u}^k\right)_{ij} \approx \sum_{g=1}^{n_g} \hat{\omega}_g \left[\epsilon\left(\psi_j\right):\mathcal{K}\left(\epsilon(\hat{u}^k_\mu), y_\mu\right):\epsilon\left(\psi_i\right)\right]\left(\hat{x}_g\right), \; 1 \le i, j \le n \ll N.$$

(4.2)

The corresponding linear systems are illustrated on Figs. 4.4 and 4.5. In particular, n should be much smaller than N to obtain a significant speedup despite the fact that the linear system is sparse in the HFM and dense in the ROM, see Sect. 2.3.6.

Based on Figs. 4.4 and 4.5, the speedup is clear for the resolution part of the linear system. We still need to illustrate how precomputations in the offline phase allow an efficient assembling of the reduced problem. Denote d the number of unknowns of second-order tensor dual quantities. In 3D, $d = 6$, for instance for the strain tensor, the unknowns are $\epsilon_{11}, \epsilon_{22}, \epsilon_{33}, \epsilon_{12}, \epsilon_{23}, \epsilon_{31}$. The reduced global tangent operator at each Newton iteration, using the reduced quadrature, can be written

$$\frac{D\mathcal{F}_\mu}{D\hat{u}}\left(\hat{u}^k\right)_{ij} \approx$$

$$\sum_{g=1}^{n_g} \hat{\omega}_g \sum_{l=1}^{d} \left[\epsilon_l\left(\psi_j\right)\left(\hat{x}_g\right)\right] \sum_{m=1}^{d} \left[\mathcal{K}_{l,m}\left(\epsilon(\hat{u}^k_\mu), y_\mu\right)\left(\hat{x}_g\right)\right]\left[\epsilon_m\left(\psi_i\right)\left(\hat{x}_g\right)\right], \; 1 \le i, j \le n.$$

(4.3)

In genericROM, the code for computing this quantity in the online stage is:

```
reducedTangentMatrix = np.einsum('g,lgj,glm,mgi->ij',
    reducedIntegrationWeights,
    reducedEpsilonAtReducedIntegPoints, localTgtMat,
    reducedEpsilonAtReducedIntegPoints, optimize = True)
```

where

- `reducedIntegrationWeights[g]` $= \hat{\omega}_g \in \mathbb{R}^{n_g}$,
- `reducedEpsilonAtReducedIntegPoints[lgj]` $= \quad \epsilon_l \left(\psi_j \right) \left(\hat{x}_g \right) \in \mathbb{R}^{d \times n_g}$,
- `localTgtMat[glm]` $= \mathcal{K}_{l,m} \left(\epsilon(\hat{u}_\mu^k), y_\mu \right) \left(\hat{x}_g \right) \in \mathbb{R}^{n_g \times d \times d}$.

The object `reducedEpsilonAtReducedIntegPoints` is a third-order tensor containing the components of the strain tensor applied to the ROB and evaluated at the reduced quadrature points. This quantity is precomputed in the offline stage. The object `localTgtMat` is a third-order tensor containing the components of the local tangent matrix evaluated at the reduced quadrature points: this quantity is computed online by the constitutive law solver.

4.2.4 Tutorials and Datasets

Various tutorials detailing some physical uses cases and capabilities of genericROM are available online.[9] Namely, variants of the POD-ECM methods are presented for nonlinear structural mechanics and nonlinear transient thermal problems, featuring various boundary conditions. The datasets needed to run these tutorials are also provided.

We detail here one tutorial,[10] corresponding to a nonlinear structural mechanics case, as described in [1]. To run this tutorial, a Z-set[11] license is required for the constitutive law solver (Z-mat).

4.2.4.1 Description of the Physics Problem

Consider a 1 m-wide cube, illustrated in Fig. 4.6. This cube is subjected to

- a variable thermal loading, in the form of time- and space-dependent scalar field obtained from a previous thermal computation, see Fig. 4.6 (right),
- a centrifugal effect generated by a rotation along the Z-axis,
- a time- and space-dependent pressure filed applied on the X_1 face,
- fixed displacement imposed along the X-axis on the X_0 face, along the Y-axis on the $X_0 Y_0$ edge and along the Z-axis on the $X_0 Y_1 Z_1$ vertex.

The modeling hypothesis are: quasistatic equilibrium equations for the deformable solid in small perturbations. The material is elastoviscoplastic with nonlinear hardening and a Norton flow. The constitutive law is specified by the .mat file.

---

[9] https://genericrom.readthedocs.io/en/latest/_tutorials/index.html.

[10] https://genericrom.readthedocs.io/en/latest/_tutorials/Mechanical/MecaSequential/index.html.

[11] http://www.zset-software.com.

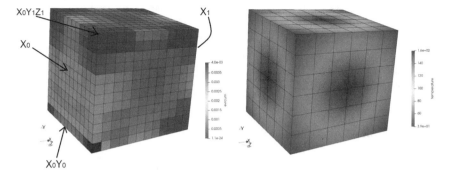

Fig. 4.6 (left) Dual mesh of the test case with the accumulated plasticity filed at the end of the simulation, (right) mesh and maximal temperature loading field (genericROM documentation)

4.2.4.2 Reduction Strategy

This is a simple "reproducting use case", where a reduced-order model is constructed to reproduce a high-fidelity solution and check its quality. This use case contains the modeling complexity of the high-pressure turbine blade cyclic extrapolation, except for the parallel computation in distributed memory and the fact that the turbine features two material (elastic and elastoviscoplastic). The number of simulation cycles can be seen as the input, while the outputs are the displacement field "U" and the accumulated plastic strain field p (named "evrcum").

The reduced-order basis is generated by the snapshot-POD method "Compress-Data" from the "DataCompressors.FusedSnapshotPOD" module.

The operator compression step is computed using the ECM (Empirical Cubature Method) "CompressOperator" from the "OperatorCompressors.Mechanical" module.

4.2.4.3 Algorithm

Snapshots are generated by Z-set and read from the Z-set format. The quality of the data compression is evaluated by computed the relative projection error of the high-fidelity solutions on the reduced-order basis:

$$\frac{\left\| u - \sum_{i=1}^{N} \left(u, \Psi_i^u \right) \Psi_i^u \right\|}{\|u\|} \leq 1 \times 10^{-5}, \qquad (4.4)$$

where Ψ_i^u denote the POD modes (or vectors of the reduced-order basis). The same quantity is considered from p.

The quality of the complete procedure is evaluated by compute the relative ℓ_2-norm error between the high-fidelity solutions and the reduced ones:

$$\frac{\left\| u - \sum_{i=1}^{N} \gamma_i^u \Psi_i^u \right\|}{\|u\|} \le 1 \times 10^{-5}, \qquad (4.5)$$

with γ_i^u the coefficients of the reduced solution (or "reducedCoordinates") computed by the reduced-order model.

4.2.4.4   Code—Offline Stage

List of imports required for the offline stage of this example:

```
from genericROM.IO import ZsetMeshReader as ZMR
from genericROM.IO import ZsetSolutionReader as ZSR
from Mordicus.Containers import ProblemData as PD
from Mordicus.Containers import CollectionProblemData as CPD
from Mordicus.Containers import Solution as S
from genericROM.FE import FETools as FT
from genericROM.DataCompressors import FusedSnapshotPOD as SP
from genericROM.OperatorCompressors import Mechanical
from Mordicus.IO import StateIO as SIO
import numpy as np
```

Then, filename and dimensions related to the mesh and the solutions have to be declared, readers are initalized, and the mesh is read:

```
folder =
    GetTestDataPath()+"Zset"+os.sep+"MecaSequential"+os.sep

meshFileName = folder + "cube.geof"
solutionFileName = folder + "cube.ut"
meshReader = ZMR.ZsetMeshReader(meshFileName)
solutionReader = ZSR.ZsetSolutionReader(solutionFileName)
mesh = meshReader.ReadMesh()
numberOfNodes = mesh.GetNumberOfNodes()
numberOfIntegrationPoints =
    FT.ComputeNumberOfIntegrationPoints(mesh)
nbeOfComponentsPrimal = 3
nbeOfComponentsDual = 6
```

Then, the part of the ECM algorithm depending only on the mesh is carried out:

```
operatorPreCompressionData =
    Mechanical.PreCompressOperator(mesh)
```

Then, the objects "Solution" are declared and populated with data from the pre-computed Z-set solutions:

```
outputTimeSequence = solutionReader.ReadTimeSequenceFromSolutionFile()
solutionU = S.Solution("U", nbeOfComponentsPrimal, numberOfNodes, primality
    = True)
```

```
solutionSigma = S.Solution("sigma", nbeOfComponentsDual,
    numberOfIntegrationPoints, primality = False)
solutionEvrcum = S.Solution("evrcum", 1, numberOfIntegrationPoints,
    primality = False)
for time in outputTimeSequence:
  solutionU.AddSnapshot(solutionReader.ReadSnapshot("U", time,
      nbeOfComponentsPrimal, primality=True), time)
  solutionSigma.AddSnapshot(solutionReader.ReadSnapshot("sig", time,
      nbeOfComponentsDual, primality=False), time)
  solutionEvrcum.AddSnapshot(solutionReader.ReadSnapshotComponent("evrcum",
      time, primality=False), time)
```

Then, the objects "CollectionProblemData" and "ProblemData" are declared, which will enable to agregate the "Solution" objects previously constructed in a standard fashion in Mordicus:

```
problemData = PD.ProblemData("MecaSequential")
problemData.AddSolution(solutionU)
problemData.AddSolution(solutionSigma)
problemData.AddSolution(solutionEvrcum)
collectionProblemData = CPD.CollectionProblemData()
collectionProblemData.AddVariabilityAxis('config', str,
    description="dummy variability")
collectionProblemData.DefineQuantity("U", "displacement", "m")
collectionProblemData.DefineQuantity("sigma", "stress", "Pa")
collectionProblemData.DefineQuantity("evrcum", "accumulated
    plasticity", "")
collectionProblemData.AddProblemData(problemData,
    config="case-1")
```

Then, the $L_2(\Omega)$ correlation operator between snapshots is computed (identified by "U"):

```
snapshotCorrelationOperator =
    {"U":FT.ComputeL2ScalarProducMatrix(mesh, 3)}
```

Then, using the snapshots-POD method, we compute the reduced-order basis for the solutions "U" with the $L_2(\Omega)$ correlation operator, and for the solutions p without correlation operator (the default operator is the identity) as a precomputing step for the Gappy-POD reconstruction method on p.

```
SP.CompressData(collectionProblemData, "U", 1.e-6,
    snapshotCorrelationOperator["U"])
SP.CompressData(collectionProblemData, "evrcum", 1.e-6)
```

Then, we compute the reduced coefficients (or "reducedCoordinates") by projecting the high-fidelity snapshots onto the reduced-order basis:

```
collectionProblemData.CompressSolutions("U",
    snapshotCorrelationOperator["U"])
```

Notice that the two previous steps can be done in one by setting the attribute "compressSolutions = True" in the function "SP.CompressData". Then, we verify the quality of the data compression on "U":

```
reducedOrderBasisU = collectionProblemData.GetReducedOrderBasis("U")
CompressedSolutionU = solutionU.GetReducedCoordinates()
compressionErrors = []
for t in outputTimeSequence:
    reconstructedCompressedSolution = np.dot(CompressedSolutionU[t],
        reducedOrderBasisU)
    exactSolution = solutionU.GetSnapshot(t)
    norml2ExSol = np.linalg.norm(exactSolution)
    if norml2ExSol != 0:
        relError =
            np.linalg.norm(reconstructedCompressedSolution-exactSolution)
        /norml2ExSol
    else:
        relError =
            np.linalg.norm(reconstructedCompressedSolution-exactSolution)
        compressionErrors.append(relError)
```

Then, we carry out the ECM algorithm to determine the reduced quadrature scheme:

```
Mechanical.CompressOperator(collectionProblemData,
    operatorPreCompressionData, mesh, 1.e-5,
listNameDualVarOutput = ["evrcum"],
    listNameDualVarGappyIndicesforECM = ["evrcum"])
```

Finally, at the end of the offline, the Modicus datamodel containing the results of this stage, is saved on disk in order to use it during the online stage.

```
SIO.SaveState("collectionProblemData", collectionProblemData)
SIO.SaveState("snapshotCorrelationOperator",
    snapshotCorrelationOperator)
```

4.2.4.5   Code—Online Stage

List of imports required for the offline stage of this example:

```
from genericROM.IO import ZsetInputReader as ZIR
from genericROM.IO import ZsetMeshReader as ZMR
from genericROM.IO import ZsetSolutionReader as ZSR
from genericROM.IO import ZsetSolutionWriter as ZSW
from Mordicus.Containers import ProblemData as PD
from Mordicus.Containers import Solution as S
from genericROM.FE import FETools as FT
from genericROM.OperatorCompressors import Mechanical as Meca
from Mordicus.IO import StateIO as SIO
import numpy as np
```

First, data saved on disk at the end of the offline stage is read:

```
collectionProblemData = SIO.LoadState("collectionProblemData")
operatorCompressionDataMechanical =
    collectionProblemData.GetOperatorCompressionData("U")
snapshotCorrelationOperator =
    SIO.LoadState("snapshotCorrelationOperator")
reducedOrderBases =
    collectionProblemData.GetReducedOrderBases()
```

Then, filename and dimensions related to the mesh and the solutions have to be declared and readers are initialized, in the same fashion as the offline stage. We mention here that the physical setting for the online stage has been taken identical to the one used in the offline stage (the folder "MecaSequential/"). In meaningful workflow, the physical setting for the online stage would be different.

```
folder =
    GetTestDataPath()+"Zset"+os.sep+"MecaSequential"+os.sep
inputFileName = folder + "cube.inp"
inputReader = ZIR.ZsetInputReader(inputFileName)
meshFileName = folder + "cube.geof"
```

Then, the mesh is read (which is required when the variability is not parametrized):

```
mesh = ZMR.ReadMesh(meshFileName)
```

Then, an object "ProblemData" is defined, which will store the data computed during the online stage:

```
onlineProblemData = PD.ProblemData("Online")
onlineProblemData.SetDataFolder(os.path.relpath(folder,
    folderHandler.scriptFolder))
```

Then, the temporal sequence and the constitutive law are read from the Z-Set input file. These are "fixed data" for the online resolution:

```
timeSequence = inputReader.ReadInputTimeSequence()
constitutiveLawsList =
    inputReader.ConstructConstitutiveLawsList()
onlineProblemData.AddConstitutiveLaw(constitutiveLawsList)
```

Then, the loadings and initial condition are read from the Z-Set input file and are reduced by projecting them onto the reduced-order basis:

```
loadingList = inputReader.ConstructLoadingsList()
onlineProblemData.AddLoading(loadingList)
for loading in onlineProblemData.GetLoadingsForSolution("U"):
    loading.ReduceLoading(mesh, onlineProblemData,
        reducedOrderBases, operatorCompressionData)
```

```
initialCondition = inputReader.ConstructInitialCondition()
onlineProblemData.SetInitialCondition(initialCondition)
initialCondition.ReduceInitialSnapshot(reducedOrderBases,
    snapshotCorrelationOperator)
```

Then, the reduced solution is computed in a nonintrusive fashion using a reduced
Newton iterative algorithm for solving the reduced nonlinear system of equations at
each time-step:

```
onlineCompressedSolution =
    Meca.ComputeOnline(onlineProblemData, timeSequence,
    operatorCompressionDataMechanical, 1.e-8)
```

Then, the reduced coefficients (or "reducedCoordinates") for the dual quantified
of interest, here p, are computed using the online part of the Gappy-POD:

```
onlineData = onlineProblemData.GetOnlineData("U")
onlineEvrcumCompressedSolution, errorGappy =
    Meca.ReconstructDualQuantity("evrcum",
    operatorCompressionDataMechanical,
onlineData, timeSequence =
    list(onlineCompressedSolution.keys()))
```

In order to compare the reduced solutions to the high-fidelity reference ones,
"Solution" objects are created and populated with precomputed Z-set solutions:

```
numberOfIntegrationPoints =
    FT.ComputeNumberOfIntegrationPoints(mesh)
nbeOfComponentsPrimal = 3
numberOfNodes = mesh.GetNumberOfNodes()
solutionFileName = folder + "cube.ut"
solutionReader = ZSR.ZsetSolutionReader(solutionFileName)
outputTimeSequence =
    solutionReader.ReadTimeSequenceFromSolutionFile()

solutionEvrcumExact = S.Solution("evrcum", 1,
    numberOfIntegrationPoints, primality = False)
solutionUExact = S.Solution("U", nbeOfComponentsPrimal,
    numberOfNodes, primality = True)
for t in outputTimeSequence:
    evrcum = solutionReader.ReadSnapshotComponent("evrcum", t,
        primality=False)
    solutionEvrcumExact.AddSnapshot(evrcum, t)
    U = solutionReader.ReadSnapshot("U", t,
        nbeOfComponentsPrimal, primality=True)
    solutionUExact.AddSnapshot(U, t)
```

"Solution" objects corresponding to the reduced solutions are constructed, populated with the reduced coefficients (or "reducedCoordinates") computed by the online
stage, and reconstructed on the complete mesh:

```
solutionEvrcumApprox = S.Solution("evrcum", 1, numberOfIntegrationPoints,
    primality = False)
solutionEvrcumApprox.SetReducedCoordinates(onlineEvrcumCompressedSolution)
solutionEvrcumApprox.UncompressSnapshots(reducedOrderBases["evrcum"])
solutionUApprox = S.Solution("U", nbeOfComponentsPrimal, numberOfNodes,
    primality = True)
solutionUApprox.SetReducedCoordinates(onlineCompressedSolution)
solutionUApprox.UncompressSnapshots(reducedOrderBases["U"])
```

Then, we verify the quality of the reduced solutions "U" and "evrcum":

```
ROMErrorsU = []
ROMErrorsEvrcum = []
for t in outputTimeSequence:
    exactSolution = solutionEvrcumExact.GetSnapshotAtTime(t)
    approxSolution = solutionEvrcumApprox.GetSnapshotAtTime(t)
    norml2ExactSolution = np.linalg.norm(exactSolution)
    if norml2ExactSolution > 1.e-10:
        relError =
            np.linalg.norm(approxSolution-exactSolution)/norml2ExactSolution
    else:
        relError = np.linalg.norm(approxSolution-exactSolution)
    ROMErrorsEvrcum.append(relError)

    exactSolution = solutionUExact.GetSnapshotAtTime(t)
    approxSolution = solutionUApprox.GetSnapshotAtTime(t)
    norml2ExactSolution = np.linalg.norm(exactSolution)
    if norml2ExactSolution > 1.e-10:
        relError =
            np.linalg.norm(approxSolution-exactSolution)/norml2ExactSolution
    else:
        relError = np.linalg.norm(approxSolution-exactSolution)
    ROMErrorsU.append(relError)
```

Finally, reduced predictions for "U" and "evrcum" are exported in the Z-set format:

```
onlineProblemData.AddSolution(solutionUApprox)
onlineProblemData.AddSolution(solutionEvrcumApprox)
ZSW.WriteZsetSolution(mesh, meshFileName, "reduced",
    collectionProblemData, onlineProblemData, "U")
```

4.2.4.6 Results

A comparison between the reduced and reference high-fidelity solutions is illustrated in Fig. 4.7.

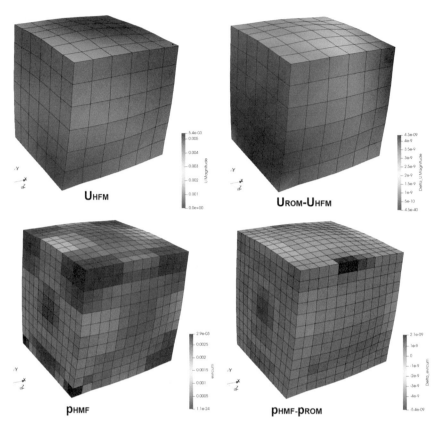

Fig. 4.7 Comparison between the reduced and reference high-fidelity solutions: (top) on the displacement "U", (bottom) on the accumulated plasticity "p" (genericROM documentation)

References

1. F. Casenave, N. Akkari, F. Bordeu, C. Rey, D. Ryckelynck, A nonintrusive distributed reduced-order modeling framework for nonlinear structural mechanics – application to elastoviscoplastic computations. Int. J. Numer. Methods Eng. **121**(1), 32–53 (2020)
2. F. Casenave, A. Gariah, C. Rey, F. Feyel, A nonintrusive reduced order model for nonlinear transient thermal problems with nonparametrized variability. Adv. Model. Simul. Eng. Sci. **7**(1), 1–19 (2020)

Open Access This chapter is licensed under the terms of the Creative Commons Attribution 4.0 International License (http://creativecommons.org/licenses/by/4.0/), which permits use, sharing, adaptation, distribution and reproduction in any medium or format, as long as you give appropriate credit to the original author(s) and the source, provide a link to the Creative Commons license and indicate if changes were made.

The images or other third party material in this chapter are included in the chapter's Creative Commons license, unless indicated otherwise in a credit line to the material. If material is not included in the chapter's Creative Commons license and your intended use is not permitted by statutory regulation or exceeds the permitted use, you will need to obtain permission directly from the copyright holder.

Chapter 5
Industrial Application: Uncertainty Quantification in Lifetime Prediction of Turbine Blades

This chapter is a synthesis of the previous ones, since many introduced concepts are applied herein. The complete ROM-net workflow, described in Sect. 2.4.2 is applied to the quantification of the uncertainty of dual quantities (such as the accumulated plastic strain and the stress tensor) on an real-life turbine blade, generated by the uncertainty of the temperature loading field. The numerical experiments make use of the codes Mordicus and genericROM, introduced respectively in Sects. 4.1 and 4.2. The content of this chapter is inspired from our publication [9].

Computing the fatigue lifetime of one such blade requires simulating its behavior until the stabilization of the mechanical response, which last several weeks using Abaqus [23] because of the size of the mesh, the complexity of the constitutive equations, and the number of loading cycles in the transient regime. With such a computation time, uncertainty quantification with the Monte Carlo method is unaffordable. In addition, such simulations are too time-consuming to be integrated in design iterations, which limits them to the final verification steps, while the design process still relies on simplified models. Accelerating these complex simulations is a key challenge while maintaining a satisfying accuracy, as it would provide useful numerical tools to improve design processes and quantify the effect of the uncertainties on the environment of the system.

Simulations are accelerated using a dictionary of reduced order models, with a classifier able to select which local reduced order model to be used for a new temperature loading. A dataset of 200 solutions is computed in a Finite Element approximation space of dimension in the order of the million, for various instances of the temperature field loading, in parallel in 7 days and 9 h on 48 cores. These solutions are computed over 11 time steps in the first cycle, using a scalable Adaptive MultiPreconditioned FETI (AMPFETI) solver [3] in Z-set finite-element software [17]. The dataset is partitioned into two clusters using a k-medoids algorithm with a ROM-oriented dissimilarity measure in 5 min; the corresponding local ROMs, using POD data compression and ECM operator compression, are trained in 2 h and 30 min. An automatic reduced model recommendation procedure, allowing to decide which local ROM to be used for a new temperature loading, is trained in the form of a logistic

© The Author(s) 2024
D. Ryckelynck et al., *Manifold Learning*, SpringerBriefs in Computer Science,
https://doi.org/10.1007/978-3-031-52764-7_5

regression classifier in 16 min. A meta-model is used to reconstruct the dual quantities of interest over the complete mesh from their values on the reduced integration points, in the form of a multi-task Lasso, which takes 1 h to train for 14 dual fields. The uncertainties on dual quantities of interest, such as the accumulated plastic strain and the stress tensor, are quantified by using our trained ROM-net on 1008 Monte Carlo draws of the temperature loading field in 2 h and 48 min, which corresponds to a speedup greater than 600 with respect to our highly optimized domain decomposition AMPFETI solver. Expected values for the Von Mises stress and the accumulated plastic strain have 0.99-confidence intervals width of respectively 1.66% and 2.84% (relative to the corresponding prediction for the expected value). As a verification stage, 20 reference solutions are computed on new temperature loadings, and dual quantities of interest are predicted with relative accuracy in the order of 1% to 2%, while the location of the maximum value is perfectly predicted.

In what follows, we describe the industrial dataset, the hypotheses of the model, and the objective of the present study. The proposed workflow for uncertainty quantification is then applied on this industrial configuration.

5.1 Industrial Context

The industrial test case of interest consists in predicting the mechanical behavior of a high-pressure (HP) turbine blade in an aircraft engine with uncertainties on the thermal loading. The industrial context and the models for the mechanical behavior and the thermal loading are presented, with a particular emphasis on the assumptions that have been made. For confidentiality reasons, mesh sizes and numerical values corresponding to the industrial dataset are not given, and the provided figures and plots do not contain any color map or physical numerical value. The accuracy of the predictions made by our methodology are given in the form of relative errors.

5.1.1 Thermomechanical Fatigue of High-Pressure Turbine Blades

High-pressure turbine blades are critical parts in an aircraft engine. Located downstream of the combustion chamber, they are subjected to extreme thermomechanical loadings resulting from the combination of centrifugal forces, pressure loads, and hot turbulent fluid flows whose temperatures are higher than the material's melting point. The repeating thermomechanical loading over time progressively damages the blades and leads to crack initiation under thermomechanical fatigue. Predicting the fatigue lifetime is crucial not only for safety reasons, but also for ecological issues, since reducing fuel consumption and improving the engine's efficiency requires increasing the temperature of the gases leaving the combustion chamber.

High-pressure turbine blades are made of monocrystalline nickel-based super-alloys that have good mechanical properties at high temperatures. To reduce the temperature inside this material, the blades contain cooling channels where flows relatively fresh air coming from the compressor. In addition, the blade's outer surface is protected by a thin thermal barrier coating. In spite of these advanced cooling technologies, the rotor blades undergo centrifugal forces at high temperatures, causing inelastic strains. Under this cyclic thermomechanical loading repeated over the flights, the structure has a viscoplastic behavior and reaches a viscoplastic stabilized response, where the dissipated energy per cycle still has a nonzero value. This is called *plastic shakedown*, and leads to *low-cycle fatigue*. At cruise flight, the persistent centrifugal force applied at high temperature induces progressive (or time-dependent) inelastic deformations: this phenomenon is called *creep*. In addition, the difference between gas pressures on the extrados and the intrados of the blade generates bending effects. Environmental factors may also locally modify the chemical composition of the material, leading to its *oxidation*. As oxidized parts are more brittle, they facilitate crack initiation and growth. *Thermal fatigue* resulting from temperature gradients is another life-limiting factor. Temperature gradients make colder parts of the structure prevent the thermal expansion of hotter parts, creating thermal stresses. Due to their higher temperatures, the hot parts are more viscous and have a lower yield stress, which make them prone to develop inelastic strains in compression. When the temperature cools down after landing, tensile *residual stresses* appear in parts which were compressed at high temperatures and favor crack nucleation. Given the complex temperature field resulting from the internal cooling channels and the turbulent gas flow, thermal fatigue has a strong influence on the turbine blade's lifetime. In particular, during transient regimes such as take-off, an important temperature gradient appears between the leading edge and the trailing edge of the blade, since the latter has a low thermal inertia due to its small thickness and thus warms up faster.

In short, the behavior of a high pressure turbine blade results from a complex interaction between low-cycle fatigue, thermal fatigue, creep, and oxidation. Due to the cost and the complexity of experiments on parts of an aircraft engine, numerical simulations play a major role in the design of high-pressure turbine blades and their fatigue lifetime assessment. All this knowledge have been learned by scientist and engineers during last decades. In the proposed approach to machine learning for model order reduction, all this knowledge is preserved in local ROMs. It is even more than that, the uncertainty propagation comes to complete this valuable traditional knowledge. We do not expect from artificial intelligence to learn everything in our modeling process.

5.1.2 Industrial Dataset and Objectives

Figure 5.1 gives the geometry and the finite-element mesh of a real high-pressure turbine blade. The mesh is made of quadratic tetrahedral elements, and contains a number of nodes in the order of the million. The elasto-viscoplastic mechani-

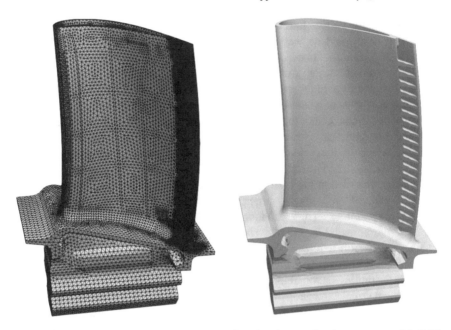

Fig. 5.1 High-pressure turbine blade geometry and mesh (micro-perforations are not modeled) [9]

cal behavior is described by a crystal plasticity model presented in the appendix of [9]. As explained above, Monte Carlo simulations using a commercial software as Abaqus are unaffordable. With the help of domain decomposition methods, the computation time can be reduced by solving equilibrium equations in parallel on different subdomains of the geometry. Using the implementation of the Adaptive MultiPreconditioned FETI solver [3] in Z-set finite-element software [17], the simulation of one single loading cycle of the HP turbine blade with 48 subdomains takes approximately 53 min.

The objective is to use a ROM-net to quantify uncertainties on the mechanical behavior of the high-pressure turbine blade, given uncertainties on the thermal loading. The reduction of the computation time should enable Monte Carlo simulations for uncertainty quantification. In particular, we are not interested in predicting the state of the structure after a large number of flight-representative loading cycles. Only one cycle is simulated. Cyclic extrapolation of the behavior of a high-pressure turbine blade has been studied in [4] and is out of the scope of this section.

5.1.3 Modeling Assumptions

It is assumed that the heat produced or dissipated by mechanical phenomena has negligible effects in comparison with thermal conduction, which enables avoiding

Fig. 5.2 Function $\omega(t)$ defining one cycle for the rotation speed [9]

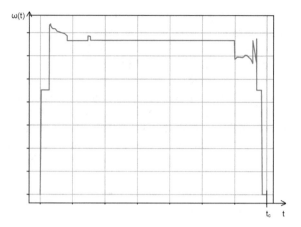

strongly coupled thermomechanical simulations and running thermal and mechanical simulations separately instead. Under a weak thermomechanical coupling, the first step consists in solving the heat equation to determine the temperature field and its evolution over time. The temperature field history defines the thermal loading and is used to compute thermal strains and temperature-dependent material parameters for the mechanical constitutive laws. Once the thermal loading is known, the temperature-dependent mechanical problem must be solved in order to predict the mechanical response of the structure.

The thermomechanical loading applied to the high-pressure turbine blade during its whole life is modeled as a cyclic loading, with one cycle being equivalent to one flight. The rotation speed of the turbine's rotor is proportional to a periodic function of time $\omega(t)$ whose evolution over one period (or cycle, see Fig. 5.2) is representative of one flight with its three main regimes, namely take-off, cruise, and landing. The period (or duration of one cycle) is denoted by t_c. The rotation speed between flights k and $k + 1$ is zero, which means that $\omega(kt_c) = 0$ for any integer k. The rotation speed $\omega(t)$ is scaled so that its maximum is 1.

Let $\Omega \subset \mathbb{R}^3$ denote the solid body representing the high-pressure turbine blade, with $\partial\Omega$ denoting its outer surface. Let $\partial\Omega^P \subset \partial\Omega$ be the surface corresponding to the intrados and extrados. The thermal loading is defined as:

$$\forall \boldsymbol{\xi} \in \Omega, \quad \forall t \in \mathbb{R}_+, \quad T(\boldsymbol{\xi}, t) = (1 - \omega(t))T_0 + \omega(t)T_{\max}(\boldsymbol{\xi}), \tag{5.1}$$

where $T_0 = 293$ K and $T_{\max}$ is the temperature field obtained when the rotation speed reaches its maximum. This field $T_{\max}$ is obtained either by an aerothermal simulation or by a stochastic model, as explained later. Similarly, the pressure load applied on $\partial\Omega^P$ reads:

$$\forall \boldsymbol{\xi} \in \partial\Omega^P, \quad \forall t \in \mathbb{R}_+, \quad p^{\partial\Omega}(\boldsymbol{\xi}, t) = (1 - \omega(t))p_0^{\partial\Omega} + \omega(t)p_{\max}^{\partial\Omega}(\boldsymbol{\xi}), \tag{5.2}$$

where $p_0^{\partial\Omega} = 1$ atm is the atmospheric pressure at sea level, and where $p_{\max}^{\partial\Omega}$ is the pressure field obtained when the rotation speed reaches its maximum. The clamping of the blade's fir-tree foot on the rotor disk is modeled by displacements boundary conditions that are not detailed here.

Small geometric details of the structure have been removed to simplify the geometry. Nonetheless, the main cooling channels are considered. The effects of the thermal barrier coating (TBC) have been integrated in aerothermal simulations, but the TBC is not considered in the mechanical simulation although its damage locally increases the temperature in the nickel-based superalloy and thus affects the fatigue resistance of the structure. Additional centrifugal effects due to the TBC are not taken into account.

The predicted mechanical response of the structure depends on many different factors. Below is a nonexhaustive list of influential factors that are possible sources of uncertainties in the numerical simulation:

- **Thermal loading:** The viscoplastic behavior of the nickel-based superalloy is very sensitive to the temperature field and its gradients. However, the temperature field is not accurately known because of the impossibility of validating numerical predictions experimentally. Indeed, temperature-sensitive paints are accurate to within 50 K only, and they do not capture a real surface temperature field since they measure the maximum temperature reached locally during the experiment.
- **Crystal orientation:** Because of the complexity of the manufacturing process of monocrystalline blades, the orientation of the crystal is not perfectly controlled. As the superalloy has anisotropic mechanical properties, defaults in crystal orientation highly affect the location of damaged zones in the structure.
- **Mechanical loading:** The centrifugal forces are well known because they are related to the rotation speed that is easy to measure. On the contrary, pressure loads are uncertain because of the turbulent nature of the incoming fluid flow. However, the effects of pressure loads uncertainties on the mechanical response are less significant than those of the thermal loading and crystal orientation uncertainties.
- **Constitutive laws:** Uncertainties on the choice of the constitutive model, the relevance of the modeling assumptions, and the values of the calibrated parameters involved in the constitutive equations also influence the results of the numerical simulations.

For simplification purposes, the only source of uncertainty that is considered in this work is the thermal loading. The equations of the mechanical problem are then seen as parametrized equations, where the parameter is the temperature field $T_{\max}$ (see Eq. (5.1)) obtained when the rotation speed reaches its maximum value. The dimension of the parameter space is then the number of nodes in the finite-element mesh. The mechanical loading is assumed to be deterministic. With the crystal orientation, the constitutive laws and their parameters (or coefficients), they are considered as known data describing the context of the study and given by experts. For details on the constitutive law model, we refer to the appendix of [9].

5.1.4 Stochastic Model for the Thermal Loading

A stochastic model is required to take into account the uncertainties on the thermal loading. Given the definition of the thermal loading in Eq. (5.1), we only need to model uncertainties in space through the field $T_{\max}$ obtained when the rotation speed reaches its maximum value. The random temperature fields must satisfy some constraints: they must satisfy the heat equation, and they must not take values out of the interval $[0\ K;\ T_{\mathrm{melt}}]$, where T_{melt} is the melting point of the superalloy. These random fields are obtained by adding random fluctuations to a reference temperature field, see Fig. 5.3. The reference field and comes from aerothermal simulations run with the software *Ansys Fluent*.[1] The data-generating distribution is defined as a Gaussian mixture model made of two Gaussian distributions with the same covariance function but with distinct means, and with a prior probability of 0.5 for each Gaussian distribution. The Gaussian distributions are obtained by taking the four first eigenfunctions of the covariance function (see Karhunen-Loève expansion [16]), with a standard deviation of 15 K. Therefore, realizations of the random temperature field read:

$$\forall \boldsymbol{\xi} \in \Omega, \quad T(\boldsymbol{\xi}) = T_{\mathrm{ref}}(\boldsymbol{\xi}) + \Upsilon_0\, \delta T_0(\boldsymbol{\xi}) + \sum_{i=1}^{4} \Upsilon_i\, \delta T_i(\boldsymbol{\xi}), \qquad (5.3)$$

where T_{ref} is the reference field, δT_0 is a temperature perturbation at the trailing edge whose maximum value is 50 K, $\{\delta T_i\}_{1 \leq i \leq 4}$ are fluctuation modes, Υ_0 is a random variable following the Bernoulli distribution with parameter 0.5, and $\{\Upsilon_i\}_{1 \leq i \leq 4}$ are independent and identically distributed random variables following the standard normal distribution $\mathcal{N}(0, 1)$. The variable Υ_0 is also independent of the other variables Υ_i. The different fields involved in Eq. (5.3) can be visualized in Fig. 5.3. Equation (5.3) defines a mixture distribution with two Gaussian distributions whose means are T_{ref} and $T_{\mathrm{ref}} + \delta T_0$. We voluntarily define this mixture distribution with δT_0 adding 50 K in a critical zone of the turbine blade in order to check that our cluster analysis can successfully detect two relevant clusters, i.e., one for fields obtained with $\Upsilon_0(\theta) = 0$ and one for fields obtained with $\Upsilon_0(\theta) = 1$. Indeed, the temperature perturbation δT_0 is expected to significantly modify the mechanical response of the high-pressure turbine blade. All the fields $\{\delta T_i\}_{0 \leq i \leq 4}$ satisfy the steady heat equation like T_{ref}, which ensures that the random fields always satisfy the heat equation under the assumption of a linear thermal behavior. For nonlinear thermal behaviors, Eq. (5.3) would define surface temperature fields that would be used as Dirichlet boundary conditions for the computation of bulk temperature fields. The assumption of a linear thermal behavior is adopted here to avoid solving the heat equation for every realization of the random temperature field.

Let us now give more details about the construction of the fluctuation modes $\{\delta T_i\}_{1 \leq i \leq 4}$. First, surface fluctuation modes are computed on the boundary $\partial \Omega$ using the method given in [21] for the construction of random fields on a curved surface.

---

[1] https://www.ansys.com/products/fluids/ansys-fluent.

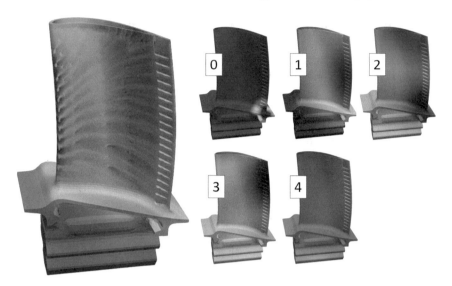

Fig. 5.3 Reference temperature field (on the left), temperature perturbation at the trailing edge (field $0 = \delta T_0$), and fluctuation modes (fields 1 to 4). The fluctuations in the fourth mode are located inside the blade, in the cooling channels [9]

The correlation function is defined as a function of the geodesic distance d_G along the surface $\partial\Omega$:

$$\rho(\xi, \xi') = \exp\left(-\frac{d_G(\xi, \xi')}{d_G^0}\right), \qquad (5.4)$$

where d_G^0 is a correlation length. Geodesic distances are computed using the algorithm described in [18, 22] and implemented in the Python library *gdist*.[2] A covariance matrix is built by evaluating the correlation function on pairs of nodes of the outer surface of the finite-element mesh, and multiplying the correlation by the constant variance. The four surface modes are then obtained by finding the four eigenvectors corresponding to the largest eigenvalues of the covariance matrix. The steady heat equation with Dirichlet boundary conditions is solved for each of these surface modes to derive the 3D fluctuation modes, using *Z-set* [17] finite-element solver. The Python library *BasicTools*[3] developed by SafranTech is used to read the finite-element mesh and write the temperature fields in a format that can be used for simulations on *Z-set*.

---

[2] https://pypi.org/project/gdist/.

[3] https://gitlab.com/drti/basic-tools.

5.2 ROM-net Based Uncertainty Quantification Applied to an Industrial High-Pressure Turbine Blade

This section develops the different stages of the ROM-net for the industrial test case presented in the previous section. Given our budget of 200 high-fidelity simulations, a dictionary containing two local ROMs is constructed using our clustering procedure. A logistic regression classifier is trained for automatic model recommendation using information identified by feature selection, followed by an alternative to the Gappy-POD for full-field reconstruction of dual quantities. Then, the results of the uncertainty quantification procedure are presented. Finally the accuracy of the ROM-net is validated using simulations for new temperature loadings.

5.2.1 Design of Numerical Experiments

Given the computational cost of high-fidelity mechanical simulations of the high-pressure turbine blade, the training data are sampled from the stochastic model for the thermal loading using a design of experiments. Our computational budget corresponds to 200 high-fidelity simulations, so a database of 200 temperature fields must be built. This database includes two separate datasets coming from two independent DoEs:

- The first dataset is built from a Maximum Projection LHS design (*MaxProj LHS DoE* [13]) and contains 80 points. This dataset will be used for the construction of the dictionary of local ROMs via clustering. The MaxProj LHS DoE has good space-filling properties on projections onto subspaces of any dimension.
- The second dataset is built from a Sobol' sequence (*Sobol' DoE*) of 120 points. Using a suboptimal DoE method ensures that this second dataset is different and independent from the first one. The lower quality of this dataset with respect to the first one is compensated by its larger population. This dataset will be used for learning tasks requiring more training examples than the construction of the local ROMs, namely the classification task for automatic model recommendation, and the training of cluster-specific surrogate models for the reconstruction of full fields from hyper-reduced predictions on a reduced-integration domain. These surrogate models (*Gappy surrogates*) replace the Gappy-POD [11] method that is commonly used in hyper-reduced simulations to retrieve dual variables on the whole mesh.

These DoEs are built with the platform *Lagun*.[4] The fact that these two datasets come from two separate DoEs is beneficial: as each of them is supposed to have good space-filling properties, they are both representative of the possible thermal loading and can therefore be used to define a training set and a test set for a given learning task. For instance, the classifier trained on the Sobol' DoE can be tested

---

[4] https://gitlab.com/drti/lagun.

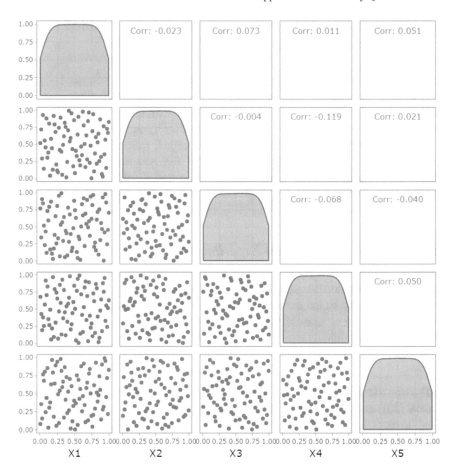

Fig. 5.4 Visualization of the MaxProj LHS DoE. The marginal distributions are represented on the diagonal. The 5D DoE is projected on 2D subspaces for visualization purposes, in order to check space-filling properties in 2D [9]

on the MaxProj LHS DoE. The local ROMs built from snapshots belonging to the MaxProj LHS DoE can make predictions on the Sobol' DoE that will be used for the training of the Gappy surrogates, which is relevant since the Gappy surrogates are supposed to analyze ROM predictions on new unseen data in the exploitation phase.

Drawing random temperature fields as defined in Eq. (5.3) requires sampling data from the random variables $\{\Upsilon_i\}_{0 \leq i \leq 4}$, where Υ_0 follows the Bernoulli distribution with parameter 0.5 and the variables Υ_i for $i \in [\![1; 4]\!]$ are independent standard normal variables and independent of Υ_0. Both DoE methods (Maximum Projection LHS and Sobol' sequence) generate point clouds with a uniform distribution in the unit hypercube. Figures 5.4 and 5.5 show the projections onto 2-dimensional subspaces of the 5D point clouds used to build our datasets. The marginal distributions are plotted to check that they well approximate the uniform distribution. These point

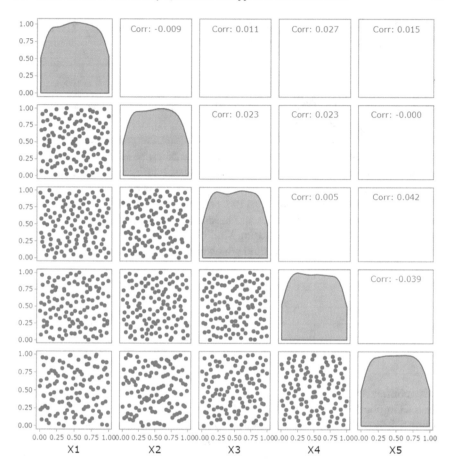

Fig. 5.5 Visualization of the Sobol' DoE. The marginal distributions are represented on the diagonal. The 5D DoE is projected on 2D subspaces for visualization purposes, in order to check space-filling properties in 2D [9]

clouds, considered as samples of a random vector $(\chi_0, \chi_1, \chi_2, \chi_3, \chi_4)$ following the uniform distribution on the unit hypercube, are transformed into realizations of the random vector $(\Upsilon_0, \Upsilon_1, \Upsilon_2, \Upsilon_3, \Upsilon_4)$ using the following transformations:

$$\Upsilon_0 = \mathbb{1}_{\chi_0 > 1/2} \quad \text{and} \quad \forall i \in [\![1; 4]\!], \quad \Upsilon_i = F^{-1}(\chi_i), \tag{5.5}$$

where F^{-1} is the inverse of the cumulative distribution function of the standard normal distribution. The resulting samples define the MaxProj dataset and the Sobol' dataset of random temperature fields, using Eq. (5.3). Each temperature field defines a thermal loading, using Eq. (5.1). The 200 corresponding mechanical problems are solved for one loading cycle with the finite-element software *Z-set* [17] with the

domain decomposition method described in [3], with 48 subdomains. The average
computation time for one simulation is 53 min.

5.2.2 ROM Dictionary Construction

The 80 simulations associated to the MaxProj dataset are used as clustering data.
Loading all the simulation data and computing the pairwise ROM-oriented dissimilarities takes about 5 min. The ROM-oriented dissimilarity defined in [7, Definition 3.11] is computed with $n = 1$, i.e., each simulation is represented by one field.
The dataset is partitioned into two clusters using our implementation of PAM [14, 15]
k-medoids algorithm, with 10 different random initializations for the medoids. The
clustering results can be visualized using Multidimensional Scaling (MDS) [2]. MDS
is an information visualization method which consists in finding a low-dimensional
dataset $\mathbf{Z}_0$ whose matrix of Euclidean distances $\mathbf{d}(\mathbf{Z}_0)$ is an approximation of the true
dissimilarity matrix δ. To that end, a cost function called stress function is minimized
with respect to $\mathbf{Z}$:

$$\mathbf{Z}_0 = \arg \min_{\mathbf{Z}} (\varsigma(\mathbf{Z}; \delta)) = \arg \min_{\mathbf{Z}} \left(\sum_{i<j} (\delta_{ij} - d_{ij}(\mathbf{Z}))^2 \right). \tag{5.6}$$

This minimization problem is solved with the algorithm Scaling by MAjoriz-
ing a COmplicated Function (SMACOF, [10]) implemented in Scikit-Learn [19].
Figure 5.6 show the clusters on the MDS representations with the goal-oriented variants of the ROM-oriented dissimilarity measure, applied on the accumulated plastic
strain p^o_{cum}. The clustering results is compared with the expected clusters corresponding to $\Upsilon_0 = 0$ and $\Upsilon_0 = 1$, the latter corresponds to the perturbation δT_0 being
activated. The obtained clusters almost correspond to the expected ones, with only 4
points with wrong labels out of 80, which quantifies the ability of the ROM-oriented
dissimilarity measure on the accumulated plasticity to infer the correct value of Υ_0.
The medoids of the two clusters are given in Fig. 5.7. Cluster 0 contains temperature
fields for which $\Upsilon_0 = 1$, while cluster 1 contains fields for which $\Upsilon_0 = 0$. It can be
observed that the quantity of interest clearly differs from one cluster to the other,
while the differences are hardly visible on the displacement field. The displacement
field combines deformations associated to different phenomena (thermal expansion,
elastic strains, viscoplastic strains) that are not necessarily related to damage in the
structure, which could explain why the quantity of interest p^o_{cum} seems to be more
appropriate for clustering in this example.

   The simulations used for the clustering procedure can directly provide snapshots
for the construction of the local ROMs. To control the duration of their training,
only 20 simulations are selected to provide snapshots for the each local ROMs,
which represents half of the clusters' populations. These simulations are selected in
a maximin greedy approach starting from the medoid (see [8, Algorithm 2, Stage 2]

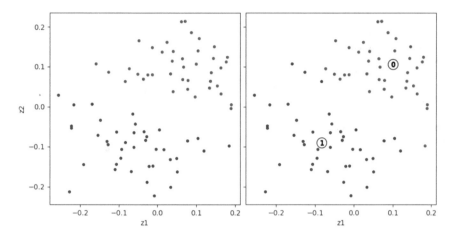

Fig. 5.6 MDS representation of the clustering results using the ROM-oriented dissimilarity measure on the quantity of interest p_{cum}^o (goal-oriented variant). On the left, the colors correspond to the expected clusters. On the right, the colors correspond to the clusters identified by the clustering algorithm. The positions of the labels 0 and 1 coincide with the positions of the clusters' medoids. The MDS relative error $\varsigma(\mathbf{Z}_0; \boldsymbol{\delta})/\varsigma(\mathbf{0}; \boldsymbol{\delta})$ is 12% [9]

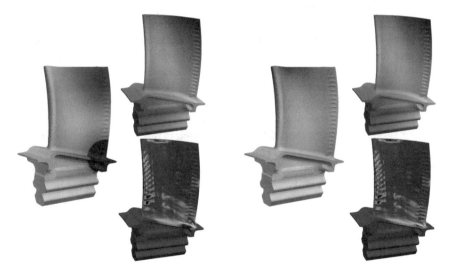

Fig. 5.7 The 3 fields on the left correspond to the medoid of cluster 0, and those on the right correspond to the medoid of cluster 1. The fields in the first and the third columns show the differences between the medoids' temperature fields and the reference temperature field T_{ref} (the scale is truncated for the first field). The second and the fourth columns show the displacement magnitude field $\sqrt{\boldsymbol{u}.\boldsymbol{u}}$ (top) and the quantity of interest p_{cum}^o (bottom) [9]

Fig. 5.8 MDS representation of the clustering results. Orange points represent the snapshots selected for cluster 0, while the light blue points represent the snapshots selected for cluster 1. For each cluster, the snapshots are selected by a maximin procedure starting from the medoid [9]

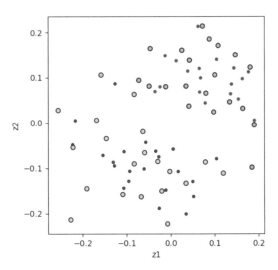

for a example of maximin selection). Figure 5.8 shows which simulations have been selected for the construction of the local ROMs.

The local ROMs are built following the methodology described in Sect. 2.3. The snapshot-POD and the ECM are done in parallel with shared memory on 24 cores. The tolerance for the snapshot-POD is set to 10^{-8} for the displacement field, and to 10^{-4} for dual variables (the quantity of interest p^o_{cum} and the six components of the stress tensor). The POD bases for the dual variables will be used for their reconstruction with the Gappy surrogates. The tolerance for the ECM is set to 5×10^{-4}. Locals ROMs each contain 18 displacement modes and between 8 and 13 modes for stress components. The first and second local ROMs contain respectively 10 and 12 modes for the quantity of interest p^o_{cum}. The ECM selects 506 (resp. 510) integration points for the reduced-integration domain of ROM 0 (resp. 1). Building one local ROM takes approximately 2 h and 30 min.

5.2.3   Automatic Model Recommendation

In this section, a classifier is trained for the automatic model recommendation task. The 120 temperature fields coming from the Sobol' dataset are used as training data for the classifier. Their labels are determined by finding their closest medoid in terms of the ROM-oriented dissimilarity measure. Hence, for each temperature field of the Sobol' dataset, two dissimilarities are computed: one with the medoid of the first cluster, and one with the medoid of the second cluster. Once trained, the classifier can be evaluated on the 80 labelled temperature fields of the MaxProj dataset.

Each temperature field is discretized on the finite-element mesh, which contains in the order of the million nodes. To reduce the dimension of the input space and facil-

Fig. 5.9 Feature selection results. The kriging metamodel for redundancy terms is represented by the red curve and built from 800 true redundancy terms (blue points). The elements containing the selected nodes are represented in the turbine blade geometry [9]

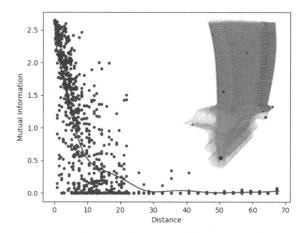

itate the training phase of the classifier, we apply the geostatistical mRMR feature selection algorithm described in [8, Algorithm 1] on data from the Sobol' dataset. First, 800 pairs of nodes are selected in the mesh, which takes 18 s. The 800 corresponding redundancy terms are computed with Scikit-Learn [19] in less than 3 s. Figure 5.9 plots the values of these redundancy terms versus the Euclidean distance between the nodes. We observe that the correlation between the redundancy mutual information terms and the distance between the nodes is poor, with a lot of noise. This can be due to the fact that the random temperature fields have been built using Gaussian random fields on the outer surface with an isotropic correlation function depending on the geodesic distance along the surface rather than the Euclidean distance. Since the turbine blade is a relatively thin structure, two nodes, one on the intrados and another one on the extrados, can be close to each other in the Euclidean distance, but with totally uncorrelated temperature fluctuations because of the large geodesic distance separating them. On the contrary, two points on the same side of the turbine blade can have correlated temperature variations while being separated by a Euclidean distance in the order of the blade's thickness. The length of the mutual information's high-variance regime seems to correspond to the blade's chord, which supports this explanation. The thinness of the turbine blade induces anisotropy in the correlation function of the bulk Gaussian random field defining the thermal loading, which implies an anisotropic behavior of the mutual information according to [8, Property 1]. The use of a local temperature perturbation δT_0 in conjunction with fluctuation modes having larger length scales may also partially explain the large variance of redundancy terms. Nonetheless, it remains clear that redundancy terms are smaller as for large distances. This trend is captured by a kriging metamodel (Gaussian process regression) trained with Scikit-Learn in a few seconds, with a sum-kernel involving the Matérn kernel with parameter 5/2 (to get a continuous and twice differentiable metamodel) and length scale 1, and a white kernel to estimate the noise level of the signal. The curve of the metamodel is given in Fig. 5.9. Then, for each node of the finite-element mesh, the mutual information with the label vari-

Table 5.1 Classification results

Class	Precision	Recall	F1-score	Support
0	0.9744	1.0000	0.9870	38
1	1.0000	0.9762	0.9880	42
Accuracy	–	–	0.9875	80
Macro avg	0.9872	0.9881	0.9875	80
Weighted avg	0.9878	0.9875	0.9875	80

able is computed. The computations of these relevance terms (in the order of the million terms) are distributed between 280 cores, which gives a total computation time of 15 min. Among these features, 5, 986 features are preselected by discarding those with a relevance mutual information lower than 0.05. The geostatistical mRMR selects 11 features in 42 s. The corresponding nodes in the finite-element mesh can be visualized in Fig. 5.9.

Remark 5.1 The metamodel for redundancy terms could be improved by defining it as a function of the precomputed geodesic distances along the outer surface rather than the Euclidean distances. Each finite-element node would be associated to its nearest neighbor on the outer surface before computing the approximate mutual information from geodesic distances.

The classifier is trained on the Sobol' dataset, using the values of the temperature fields at the 11 nodes identified by the feature selection algorithm. The classifier is a logistic regression [1, 5, 6] with elastic net regularization [24] implemented in Scikit-Learn. The two hyperparameters involved in the elastic net regularization are calibrated using 5-fold cross-validation, giving a value of 0.001 for the inverse of the regularization strength, and 0.4 for the weight of the L^1 penalty term (and thus 0.6 for the L^2 penalty term). Due to the L^1 penalty term, the classifier only uses 5 features among the 11 input features. The classifier's accuracy, evaluated on the MaxProj dataset to use new unseen data, reaches 98.75%. The confusion matrix indicates that 100% of the test examples belonging to class 0 have been correctly labeled, and that 2.38% of the test examples belonging to class 1 have been misclassified. Table 5.1 summarizes the values of precision, recall and F1-score on test data.

5.2.4 Surrogate Model for Gappy Reconstruction

When using hyper-reduction, the ROM calls the constitutive equations solver only at the integration points belonging to the reduced-integration domain. It is recalled that the ECM selected 506 (resp. 510) integration points for the reduced-integration domain of ROM 0 (resp. 1), and that the finite-element mesh initially contains a number of integration points in the order of the million. Therefore, after a reduced simulation, dual variables defined at integration points are known only at integration points of the reduced-integration domain. To retrieve the full field, the Gappy-POD [11] finds the coefficients in the POD basis that minimize the squared error between the reconstructed field and the ROM predictions on the reduced-integration domain. This minimization problem defines the POD coefficients as a linear function of the predicted values on the reduced-integration domain. Although these coefficients are optimal in the least squares sense, they can be biased by the errors made by the ROM. To alleviate this problem, we propose to replace the common Gappy-POD procedure by a metamodel or *Gappy surrogate*. The inputs and the outputs of the Gappy surrogate are the same as for the Gappy-POD: the input is a vector containing the values of a dual variable on the reduced-integration domain, and the output is a vector containing the optimal coefficients in the POD basis. One Gappy surrogate must be built for each dual variable of interest: in our case, 7 surrogate models per cluster are required, namely one for the quantity of interest p_{cum}^o and one for every component of the Cauchy stress tensor.

The training data for these Gappy surrogates are obtained by running reduced simulations with the local ROMs, using the thermal loadings of the Sobol' dataset. Indeed, the two local ROMs have been built on the MaxProj dataset, therefore thermal loadings of the Sobol' dataset can play the role of test data for the ROMs. For each thermal loading in the Sobol' dataset, the true high-fidelity solution is already known since it has been computed to provide training data for the classifier. In addition, the exact labels for these thermal loadings are known, which means that we know which local ROM to choose for each thermal loading of the Sobol' dataset. Given ROM predictions on the reduced-integration domain, the optimal coefficients in the POD basis are given by the projections of the true prediction made by the high-fidelity model (the finite-element model) onto the POD modes. This provides the true outputs for the Sobol' dataset, which can then be used as a training set for the Gappy surrogates.

Given the high-dimensionality of the input data (there are more than 500 integration points in the reduced-integration domains) with respect to the number of training examples (120 examples), a multi-task Lasso metamodel is used. The hyperparameter controlling the regularization strength is optimized by 5-fold cross-validation. Training the 14 Gappy surrogates (7 for each cluster) takes 1 h. The Gappy surrogates select between 8% and 18% of the integration points in the reduced-integration domains, due to the L^1 regularization. The mean cross-validated coefficients of determination are 0.9637 (resp. 0.8935) for the quantity of interest for cluster 0 (resp. cluster 1), and range from 0.9404 to 0.9938 for stress components. These satisfying results mean that it is not required to train a kriging metamodel with the variables selected by Lasso to get nonlinear Gappy surrogates. The Gappy surrogates are then linear, just as the Gappy-POD.

The accuracy gains provided by the Gappy surrogates with respect to classical Gappy-POD on the present industrial case is investigated in Fig. 5.10. Here, 24 high-fidelity simulation in the first cluster are computed (with $\Upsilon_0 = 0$) as reference, and Gappy surrogates (using two meta models Lasso and ElasticNet) and classical Gappy-POD are computed using ROM 0. For both variants of meta models and

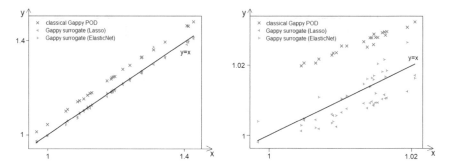

Fig. 5.10 Mean over the complete mesh of dual quantities of interest: accumulated plastic strain p_{cum}^o (left) and component 33 of the stress tensor σ_{33} (right), plotted as points where the x-coordinate is the reference value, and the y-coordinate is the considered reduced prediction [9]

both quantities of interest: accumulated plastic strain and the component 33 of the stress tensor, Gappy surrogates provides more accurate predictions than the classical Gappy-POD.

Remark 5.2 In this strategy, the local ROMs solve the equations of the mechanical problem, which enables using linear surrogate models to reconstruct dual variables. Using surrogate models from scratch instead of local ROMs would have been more difficult, given the nonlinearities of this mechanical problem and the lack of training data for regression. In addition, such surrogate models would require a parametrization of the input temperature fields, whereas the local ROMs use the exact values of the temperature fields on the RID without assuming any model for the thermal loading.

The dictionary-based ROM-net used for mechanical simulations of the high-pressure turbine blade is made of a dictionary of two local hyper-reduced order models and a logistic regression classifier. The classifier analyzes the values of the input temperature field at 11 nodes only, identified by our feature selection strategy. For a given thermal loading in the exploitation phase, after the reduced simulation with the local ROM recommended by the classifier, linear cluster-specific Gappy surrogates reconstruct the full dual fields (quantity of interest and stress components) from their predicted values on the reduced-integration domain.

5.2.5   Uncertainty Quantification Results

Once trained, the ROM-net can be applied for the quantification of uncertainties on the mechanical behavior of the HP turbine blade resulting from the uncertainties on the thermal loading. Since the ROM-net online operations can be performed sequentially on one single core, 24 cores are used in order to compute the solution for 24 thermal loadings at once. This way, 42 batches of 24 Monte Carlo simulations

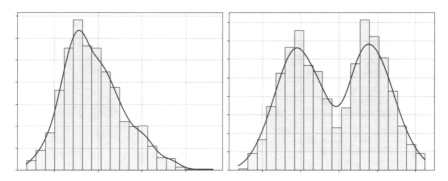

Fig. 5.11 Histograms and probability density functions of the quantities of interest $\overline{p}^o_{cum}$ (left) and $\overline{\sigma}_{eq}$ (right) [9]

Table 5.2 Widths of the confidence intervals (CI) for the expectations, expressed as percentages of the estimated expectations

Estimated variable	Confidence level	Relative CI width (%)
$e[\overline{p}^o_{cum}]$	0.95	2.16
$e[\overline{p}^o_{cum}]$	0.99	2.84
$e[\overline{\sigma}_{eq}]$	0.95	1.26
$e[\overline{\sigma}_{eq}]$	0.99	1.66

are run in 2 h and 48 min. The 1008 thermal loadings used for this study are generated by randomly sampling points from the uniform distribution on the 5D unit hypercube and applying the transformation given in Eq. (5.5).

The expected values of $\overline{p}^o_{cum}$ and $\overline{\sigma}_{eq}$ are estimated with the empirical means $\overline{Z}_n = \frac{1}{n} \sum_{i=1}^{n} Z_i$, where Z_i are the corresponding samples. The variances of $\overline{p}^o_{cum}$ and $\overline{\sigma}_{eq}$ are computed using the unbiased sample variance $S_n^2 = \frac{1}{n-1} \sum_{i=1}^{n} (Z_i - \overline{Z}_n)^2$.

The Central Limit Theorem gives asymptotic confidence intervals for the expected values: for all $\alpha \in]0; 1[$,

$$I_n = \left[\overline{Z}_n - \phi_{1-\frac{\alpha}{2}} \sqrt{S_n^2/n}; \overline{Z}_n + \phi_{1-\frac{\alpha}{2}} \sqrt{S_n^2/n}, \right], \quad (5.7)$$

where ϕ_r denotes the quantile of order r of the standard normal distribution $\mathcal{N}(0, 1)$, and I_n is an asymptotic confidence interval with confidence level $1 - \alpha$ for the expectation η: $\lim_{n \to +\infty} \mathbb{P}(\eta \in I_n) = 1 - \alpha$. The widths of the confidence intervals are expressed as a percentage of the estimated value for the expectations in Table 5.2.

The probability density functions of the quantities of interest can be estimated using Gaussian kernel density estimation (see Sect. 6.6.1. of [12]). Figure 5.11 gives

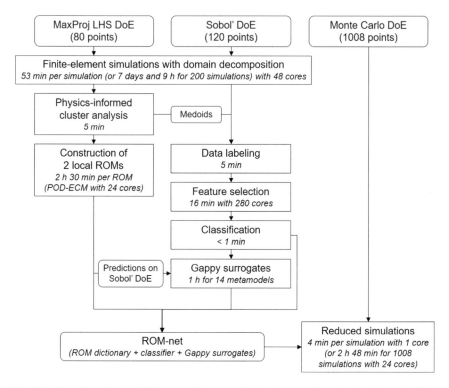

Fig. 5.12 Workflow for the ROM-net metholgy applied to the considered industrial setting [9]

the histograms and estimated distributions for $\overline{p}^{o}_{\mathrm{cum}}$ and $\overline{\sigma}_{\mathrm{eq}}$. The shapes of these distributions highly depend on the assumptions made for the stochastic thermal loading.

5.2.6   Workflow

Figure 5.12 provides an illustration of the workflow and the computational time of each step presented above.

5.2.7   Verification

For verification purposes, the accuracy of the ROM-net is evaluated on 20 Monte Carlo simulations with 20 new thermal loadings. These thermal loadings are generated by randomly sampling points from the uniform distribution on the 5D unit hypercube, and applying the transformation given in Eq. (5.5). The reduced simulations are run on single cores. The total computation time for generating a new thermal loading on the fly, selecting the corresponding reduced model, running one reduced simulation and reconstructing the quantities of interest is 4 min on average. As a comparison, one single high-fidelity simulation with Z-set [17] with 48 subdomains takes 53 min, which implies that the ROM-net computes 13.25 times faster. However, one high-fidelity simulation requires 48 cores for domain decomposition, whereas the ROM-net works on one single core. Hence, using 48 cores to run 48 reduced simulations in parallel, 636 reduced simulations can be computed in 53 min with the ROM-net, while the high-fidelity model only runs one simulation. In addition to the acceleration of numerical simulations, energy consumption is reduced by a factor of 636 in the exploitation phase. In spite of the fast development of high-performance computing, numerical methods computing approximate solutions at reduced computational resources and time are particularly important for many-query problems such as uncertainty quantification, where the intensive use of computational resources is a major concern. Model order reduction and ROM-nets play a prominent role toward *green* numerical simulations [20]. Of course, the number of simulations in the exploitation phase must be large enough to compensate the efforts made in the training phase, like in any machine learning or model order reduction problem.

Figures 5.13 and 5.14 show the results for two simulations belonging to cluster 0 and cluster 1 respectively. These figures give the difference between the current temperature field as the reference one, i.e., the field $T - T_{\text{ref}}$, and the resulting variations of the quantity of interest predicted by the ROM-net and the high-fidelity model, i.e., $p_{\text{cum}}^{o,\text{ROM}}(T) - p_{\text{cum}}^{o,\text{HF}}(T_{\text{ref}})$ and $p_{\text{cum}}^{o,\text{HF}}(T) - p_{\text{cum}}^{o,\text{HF}}(T_{\text{ref}})$. The signs and the positions of the variations of the quantity of interest seem to be quite well predicted by the ROM-net.

Let us introduce a zone of interest Ω' defined by all of the integration points at which p_{cum}^{o} is higher than $0.4 \times \max p_{\text{cum}}^{o}(\xi)$ for the thermal loading defined by $T_{\text{ref}} + \delta T_0$. This zone of interest contains 209 integration points. The values of the variables p_{cum}^{o} and σ_{eq} averaged over Ω' are denoted by $\overline{p}_{\text{cum}}^{o}$ and $\overline{\sigma}_{\text{eq}}$. Table 5.3 gives different indicators quantifying the errors made by the ROM-net: the L^2 relative errors on the whole domain Ω and on the zone of interest Ω', the L^∞ relative errors on Ω and Ω', the relative errors on $\overline{p}_{\text{cum}}^{o}$ and $\overline{\sigma}_{\text{eq}}$, and the errors on the locations of the points where the fields p_{cum}^{o} and σ_{eq} reach their maxima. All the relative errors remain in the order of 1% or 2%, which validates the methodology. In addition, the ROM-net perfectly predicts the position of the critical points at which p_{cum}^{o} and σ_{eq} reach their maxima. Figure 5.15 shows errors on the quantities of interest.

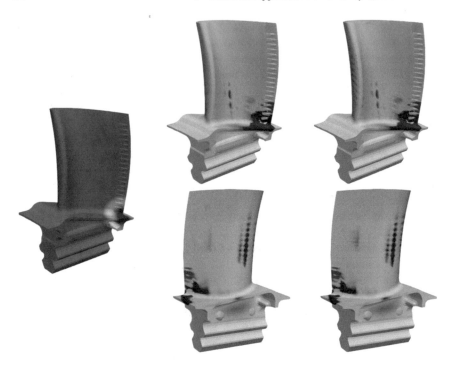

Fig. 5.13 Comparison between high-fidelity predictions (middle column) and ROM-net's predictions (right-hand column). The field on the left represents the difference between the current temperature field (belonging to cluster 0) and the reference one. The other fields correspond to the increments of the quantity of interest p_{cum}^o with respect to its reference state obtained with the reference temperature field [9]

Table 5.3 Error indicators for the evaluation of the ROM-net on 20 new thermal loadings

Error indicator	Errors on p_{cum}^o (%)	Errors on σ_{eq} (%)
Mean L^2 relative error on Ω	1.14	0.84
Mean L^2 relative error on Ω'	0.75	1.46
Mean L^∞ relative error on Ω	1.11	1.09
Mean L^∞ relative error on Ω'	1.05	2.60
Mean rel. err. on value averaged over Ω'	0.50	0.89
Mean distance between maxima	0	0

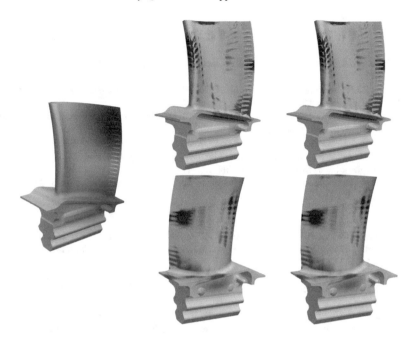

Fig. 5.14 Comparison between high-fidelity predictions (middle column) and ROM-net's predictions (right-hand column). The field on the left represents the difference between the current temperature field (belonging to cluster 1) and the reference one. The other fields correspond to the increments of the quantity of interest p_{cum}^{o} with respect to its reference state obtained with the reference temperature field [9]

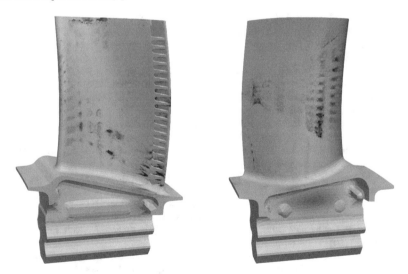

Fig. 5.15 Errors on the quantity of interest p_{cum}^{o}. The red (resp. blue) color is used for zones where the quantity of interest is overestimated (resp. underestimated) [9]

References

1. J. Berkson, Application of the logistic function to bio-assay. J. Am. Stat. Assoc. **39**(227), 357–365 (1944)
2. I. Borg, P.J.F. Groenen, *Modern Multidimensional Scaling: Theory and Applications*. (Springer Science & Business Media, 2005)
3. C. Bovet, A. Parret-Freaud, N. Spillane, P. Gosselet, Adaptive multipreconditioned FETI: Scalability results and robustness assessment. Comput. Struct. **193**, 1–20 (2017)
4. F. Casenave, N. Akkari, F. Bordeu, C. Rey, D. Ryckelynck, A nonintrusive distributed reduced-order modeling framework for nonlinear structural mechanics–application to elastoviscoplastic computations. Int. J. Numer. Methods Eng. **121**(1), 32–53 (2020)
5. D.R. Cox, The regression analysis of binary sequences. J. Roy. Stat. Soc. Ser. B (Methodol.) **20**(2), 215–242 (1958)
6. D.R. Cox, Some procedures connected with the logistic qualitative response curve, in *Research Papers in Probability and Statistics* (1966), pp. 55–71
7. T. Daniel, F. Casenave, N. Akkari, A. Ketata, D. Ryckelynck, Physics-informed cluster analysis and a priori efficiency criterion for the construction of local reduced-order bases. J. Comput. Phys. **458**, 111120 (2022)
8. T. Daniel, F. Casenave, N. Akkari, D. Ryckelynck, Data augmentation and feature selection for automatic model recommendation in computational physics. Math. Comput. Appl. **26**(1) (2021)
9. T. Daniel, F. Casenave, N. Akkari, D. Ryckelynck, C. Rey, Uncertainty quantification for industrial numerical simulation using dictionaries of reduced order models. Mech. Ind. **23**, 3 (2022)
10. J. de Leeuw, Applications of convex analysis to multidimensional scaling. in *Recent Developments in Statistics*, eds. by J.R. Barra, F. Brodeau, G. Romier, B. van Cutsem (1977), pp. 133–145
11. R. Everson, L. Sirovich, Karhunen-Loève procedure for gappy data. J. Opt. Soc. Am. A **12**(8), 1657–1664 (1995)
12. T. Hastie, R. Tibshirani, J.H. Friedman, *The Elements of Statistical Learning: Data Mining, Inference, and Prediction*, 2nd edn (Springer series in statistics, Springer, 2009)
13. V.R. Joseph, E. Gul, S. Ba, Maximum projection designs for computer experiments. Biometrika **102**(2), 3 (2015)
14. L. Kaufman, P.J.R. Leonard Kaufman, P.J. Rousseeuw, *Finding Groups in Data: An Introduction to Cluster Analysis* (A Wiley-Interscience publication, Wiley, 1990)
15. L. Kaufmann, P. Rousseeuw, Clustering by means of medoids, in *Data Analysis based on the L1-Norm and Related Methods*, vol. 1 (1987), pp. 405–416
16. B.N. Khoromskij, A. Litvinenko, H.G. Matthies, Application of hierarchical matrices for computing the Karhunen-Loève expansion. Computing **84**(1–2), 49–67 (2009)
17. M. ParisTech, ONERA, the French aerospace lab. Zset: nonlinear material & structure analysis suite. http://www.zset-software.com (1981)
18. J.S.B. Mitchell, D.M. Mount, C.H. Papadimitriou, The discrete geodesic problem. SIAM J. Comput. **16**(4), 647–668 (1987)
19. F. Pedregosa, G. Varoquaux, A. Gramfort, V. Michel, B. Thirion, O. Grisel, M. Blondel, P. Prettenhofer, R. Weiss, V. Dubourg, J. Vanderplas, A. Passos, D. Cournapeau, M. Brucher, M. Perrot, E. Duchesnay, Scikit-learn: Machine learning in Python. J. Mach. Learn. Res. **12**, 2825–2830 (2011)
20. D. Ryckelynck, D.M. Benziane, A. Musienko, G. Cailletaud, Toward "greenâŁž mechanical simulations in materials science. Eur. J. Comput. Mech. **19**(4), 365–388 (2010)
21. C. Scarth et al., *Random field simulation over curved surfaces: Applications to computational structural mechanics* (Comput. Methods Appl. Mech, Engrg, 2018)
22. V. Surazhsky, T. Surazhsky, D. Kirsanov, S.J. Gortler, H. Hoppe, Fast exact and approximate geodesics on meshes. ACM Trans. Graph. **24**(3), 553–560 (2005)

23. D. Systèmes, Abaqus Unified FEA. https://www.3ds.com/fr/produits-et-services/simulia/produits/abaqus/ (1978)
24. H. Zou, T. Hastie, Regularization and variable selection via the elastic net **67**(2), 301–320 (2005)

Open Access This chapter is licensed under the terms of the Creative Commons Attribution 4.0 International License (http://creativecommons.org/licenses/by/4.0/), which permits use, sharing, adaptation, distribution and reproduction in any medium or format, as long as you give appropriate credit to the original author(s) and the source, provide a link to the Creative Commons license and indicate if changes were made.

The images or other third party material in this chapter are included in the chapter's Creative Commons license, unless indicated otherwise in a credit line to the material. If material is not included in the chapter's Creative Commons license and your intended use is not permitted by statutory regulation or exceeds the permitted use, you will need to obtain permission directly from the copyright holder.

Chapter 6
Applications and Extensions: A Survey of Literature

This chapter contains a literature survey of the work published by the authors in the timeframe of their collaboration, where the concept presented in this book have been applied to real-life industrial settings, and new methodologies have been developed.

The listed contributions are grouped in the following themes: linear manifold learning in Sect. 6.1, nonlinear dimensionality reduction via auto-encoder in Sect. 6.2, piecewise linear dimensionality reduction via dictionary-based ROM-nets in Sect. 6.3 and manifold learning of physics problems assisted by black-box regressors in Sect. 6.4.

6.1 Linear Manifold Learning

A priori hyper-reduction method: an adaptive approach Model reduction methods are usually based on offline preliminar simulations to build the shape functions of the reduced order model (ROM) before the computation of the reduced state variables. They are a posteriori approaches. Most of the time these offline computations are as complex as the simulation which we want to simplify by the ROM. The a priori reduction method proposed in [17] avoids such preliminary computations. It is an a priori approach based on the analysis of some state evolutions, such that all the state evolutions needed to perform the model reduction are described by an approximate ROM. The ROM and the state evolution are simultaneously improved by the method, thanks to an adaptive strategy. An initial set of known shape functions can be used to define the ROM to adapt, but it is not necessary. The adaptive procedure includes extensions of the subspace spanned by the shape functions of the ROM and selections of the most relevant shape functions in order to represent the state evolution. The hyper-reduction is achieved by selecting a part of the integration points of the finite element model to forecast the evolution of the reduced state variables. In this method, both the number of degrees of freedom and the number of integration points are reduced.

© The Author(s) 2024
D. Ryckelynck et al., *Manifold Learning*, SpringerBriefs in Computer Science,
https://doi.org/10.1007/978-3-031-52764-7_6

A nonintrusive distributed reduced-order modeling framework for nonlinear structural mechanics—Application to elastoviscoplastic computations In [8] is proposed a framework that constructs reduced-order models for nonlinear structural mechanics in a nonintrusive fashion and can handle large-scale simulations. Three steps are carried out: (i) the production of high-fidelity solutions by commercial software, (ii) the offline stage of the model reduction, and (iii) the online stage where the reduced-order model is exploited. The nonintrusivity assumes that only the displacement field solution is known, and the proposed framework carries out operations on these simulation data during the offline phase. The compatibility with a new commercial code only needs the implementation of a routine converting the discretized solution into the used data format. The nonintrusive capabilities of the framework are demonstrated on numerical experiments using commercial versions of Z-set and Ansys Mechanical. The nonlinear constitutive equations are evaluated by using an external plugin. The large-scale simulations are handled using domain decomposition and parallel computing with distributed memory. The features and performances of the framework are evaluated on two numerical applications involving elastoviscoplastic materials: the second one involves a model of high-pressure blade, where the framework is used to extrapolate cyclic loadings in 6.5 h, whereas the reference high-fidelity computation would take 9.5 days.

Fast computation of transient thermal profiling of high-pressure compressors under non-parametrized boundary conditions variability In [9], a transient thermal problem is considered, with a nonlinear term coming from the radiation boundary condition and a nonparametrized variability in the form complex scenarios for the initial condition and the convection coefficients and external temperatures. A posteriori reduced order modeling by snapshot Proper Orthogonal Decomposition is used. To treat the nonlinearity, hyperreduction is required since precomputing the polynomial nonlinearities becomes too expensive for the radiation term. The Empirical Cubature Method, originally proposed for nonlinear structural mechanics, is derived for these equations. The method is applied to the design of high-pressure compressors for civilian aircraft engines, where a fast evaluation of the solution temperature is required when testing new configurations. When using in the reduced solver the same model as the one from the high-fidelity code, the approximation is very accurate. However, when using a commercial code to generate the high-fidelity data, where the implementation of the model and solver is unknown, the reduced model is less accurate but still within engineering tolerances. Hence, the regularizing property of reduced order models, together with a nonintrusive approach, enables the use of commercial software to generate the data, even under some degree of uncertainty in the proprietary model or solver of the commercial software.

Time Stable Reduced Order Modeling by an Enhanced Reduced Order Basis of the Turbulent and Incompressible 3D Navier–Stokes Equations In [3], the problem of constructing a time stable reduced order model of the 3D turbulent and incompressible Navier–Stokes equations is considered. The lack of stability associated with the order reduction methods of the Navier–Stokes equations is a well-known problem and, in general, it is very difficult to account for different scales of a turbulent flow in the same reduced space. To remedy this problem, a new stabilization technique is proposed, based on an a priori enrichment of the classical proper orthogonal decomposition (POD) modes with dissipative modes associated with the gradient of the velocity fields. The main idea is to be able to do an a priori analysis of different modes in order to arrange a POD basis in a different way, which is defined by the enforcement of the energetic dissipative modes within the first orders of the reduced order basis. This enables the modeling of the production and dissipation of the turbulent kinetic energy (TKE) in a separate fashion within the high ranked new velocity modes, hence to ensure good stability of the reduced order model. The importance of this a priori enrichment of the reduced basis is illustrated on a typical aeronautical injector with Reynolds number of 45,000. This order reduction technique is able to recover large scale features for very long integration times (25 ms in the present case). Moreover, the reduced order model exhibits periodic fluctuations with a period of 2.2 ms corresponding to the time scale of the precessing vortex core (PVC) associated with this test case.

An updated Gappy-POD to capture non-parameterized geometrical variation in fluid dynamics problems In [4], a method is proposed to fill the gap within an incomplete turbulent and incompressible data field, while satisfying the topological and intensity changes of the fluid flow after a non-parameterized geometrical variation in the fluid domain. A single baseline large eddy simulation (LES) is assumed to be performed prior geometrical variations. The method enhances the Gappy-POD method proposed by Everson and Sirovich in 1995, in the case where the given set of empirical eigenfunctions is not sufficient and is not interpolant for the recovering of the modal coefficients for each Gappy snapshot by a least squares procedure. This is typically the case when we introduce non-parameterized geometrical modifications in the fluid domain. Here, after the baseline simulation, additional solutions of the incompressible Navier–Stokes equations are solely performed over a restricted fluid domain, that contains the geometrical modifications. These local LESs, called hybrid simulations, are performed by using the immersed boundary technique, which uses of a fluid boundary condition and the baseline velocity field. Then, the POD modes are updated using a local modification of the baseline POD modes in the restricted fluid domain. Furthermore, a physical correction of the latter enhanced Gappy-POD modal coefficients is obtained thanks to a Galerkin projection of the Navier–Stokes equations upon the new modes of the available data. This enhancement procedure on the global velocity reconstruction by the physical constraint was tested on a 3D semi-industrial test case of a typical aeronautical injection system and, a 2D laminar and unsteady incompressible test case. The speed-up relative to this new technique is equal to 100, which allows the fast exploration of two new designs of the aeronautical injection system.

6.2   Nonlinear Dimensionality Reduction via Auto-Encoder

Data-Targeted Prior Distribution for Variational AutoEncoder In [1], Variational
AutoEncoders (VAE) are used to study unsteady and compressible fluid flows in aircraft engines. Inferential methods enable the computation of sharp approximations of
the posterior probability of the parameters of the VAE using the transient dynamics
of the training velocity fields, and the generatation of plausible velocity fields. An
important application is the initialization of such transient simulations. It is known
by the Bayes theorem that the choice of the prior distribution is very important for
the computation of the posterior probability, proportional to the product of likelihood
with the prior probability. A new inference model is proposed, based on a new prior
defined by the density estimate with the realizations of the kernel proper orthogonal
decomposition coefficients of the available training data. It is illustrated that this
inference model improves the results obtained with the usual standard normal prior
distribution. This new generative approach can also be seen as an improvement of
the kernel proper orthogonal decomposition method, for which we do not usually
have a robust technique for expressing the pre-image in the input physical space of
the stochastic reduced field in the feature high-dimensional space with a kernel inner
product.

**A Bayesian Nonlinear Reduced Order Modeling Using Variational AutoEn-
coders** In [2], a new nonlinear projection based model reduction using convolutional
VAEs. This framework is applied on transient incompressible flows. The accuracy
is obtained thanks to the expression of the velocity and pressure fields in a nonlinear
manifold maximising the likelihood on pre-computed data in the offline stage. A
confidence interval is obtained for each time instant thanks to the definition of the
reduced dynamic coefficients as independent random variables for which the posterior probability given the offline data is known. The parameters of the nonlinear
manifold are optimized as the ones of the decoder layers of an autoencoder. The
parameters of the conditional posterior probability of the reduced coefficients are the
ones of the encoder layers of the same autoencoder. This reduced-order model is not
a regression model over the offline pre-computed data. The numerical resolution of
the ROM is based on the Chorin projection method. This new nonlinear projection-
based reduced order modeling is applied to a 2D Karman Vortex street flow and a
3D incompressible and unsteady flow in an aeronautical injection system.

Deep multimodal autoencoder for crack criticality assessment In continuum
mechanics, the prediction of defect harmfulness requires to solve approximately
partial differential equations with given boundary conditions. In [14], boundary
conditions are learnt for tight local volumes (TLV) surrounding cracks in three-
dimensional volumes. A nonparametric data-driven approach is used to define the
space of defects, by considering defects observed via X-Ray computed tomography.
The dimension of the ambient space for the observed images of defects is huge. A
nonlinear dimensionality reduction scheme is proposed in order to train a reduced

latent space for both the morphology of defects and their local mechanical effects in the TLV. A multimodal autoencoder enables to mix morphological and mechanical data. It contains a single latent space, termed mechanical latent space. But this latent space is fed by two encoders. One is related to the images of defects and the other to mechanical fields in the TLV. The latent variables are input variables for a geometrical decoder and for a mechanical decoder. In this work, mechanical variables are displacement fields. The autoencoder on mechanical variables enables projection-based model order reduction as proposed in the study of Lee and Carlberg. The main novelty of this work is a submodeling approach assisted by artificial intelligence. Here, for defect images in the test set, Dirichlet boundary conditions are applied to TLV. These boundary conditions are forecasted by the mechanical decoder with a latent vector predicted by the morphological encoder. For that purpose, a mapping is trained to convert morphological latent variables into mechanical latent variables, denoted "direct mapping". An "inverse mapping" is also trained for error estimation with respect to morphological predictions. Errors on mechanical predictions are close to 5% with simulation speed-up ranging for 3 to 120. It is illustrated that latent variables forecasted by the images of defects are prone to a better understanding of the predictions.

6.3 Piecewise Linear Dimensionality Reduction via Dictionary-Based ROM-Nets

Model order reduction assisted by deep neural networks (ROM-net) In [12], a general framework is proposed for projection-based model order reduction assisted by deep neural networks. The proposed methodology, called ROM-net, consists in using deep learning techniques to adapt the reduced-order model to a stochastic input tensor whose nonparametrized variabilities strongly influence the quantities of interest for a given physics problem. In particular, the concept of dictionary-based ROM-nets is introduced, where deep neural networks recommend a suitable local reduced-order model from a dictionary. The dictionary of local reduced-order models is constructed from a clustering of simplified simulations enabling the identification of the subspaces in which the solutions evolve for different input tensors. The training examples are represented by points on a Grassmann manifold, on which distances are computed for clustering. This methodology is applied to an anisothermal elastoplastic problem in structural mechanics, where the damage field depends on a random temperature field. When using deep neural networks, the selection of the best reduced-order model for a given thermal loading is 60 times faster than when following the clustering procedure used in the training phase.

Mechanical dissimilarity of defects in welded joints via Grassmann manifold and machine learning Assessing the harmfulness of defects based on images is becoming more and more common in industry. At present, these defects can be inserted in digital twins that aim to replicate in a mechanical model what is observed on a component so that an image-based diagnosis can be further conducted. However, the variety of defects, the complexity of their shape, and the computational complexity of finite element models related to their digital twin make this kind of diagnosis too slow for any practical application. In [18], a classification of observed defects enables the definition of a dictionary of digital twins. These digital twins prove to be representative of model-reduction purposes while preserving an acceptable accuracy for stress prediction. Nonsupervised machine learning is used for both the classification issue and the construction of reduced digital twins. The dictionary items are medoids found by a k-medoids clustering algorithm. Medoids are assumed to be well distributed in the training dataset according to a metric or a dissimilarity measurement. A new dissimilarity measurement between defects is proposed. It is theoretically founded according to approximation errors in hyper-reduced predictions. In doing so, defect classes are defined according to their mechanical effect and not directly according to their morphology. In practice, each defect in the training dataset is encoded as a point on a Grassmann manifold. This methodology is evaluated through a test set of observed defects totally different from the training dataset of defects used to compute the dictionary of digital twins. The most appropriate item in the dictionary for model reduction is selected according to an error indicator related to the hyper-reduced prediction of stresses. No plasticity effect is considered here (merely isotropic elastic materials), which is a strong assumption but which is not critical for the purpose of this work. In spite of the large variety of defects, accurate predictions of stresses for most of defects in the test set are obtained.

Multimodal data augmentation for digital twining assisted by artificial intelligence in mechanics of materials Digital twins in the mechanics of materials usually involve multimodal data in the sense that an instance of a mechanical component has both experimental and simulated data. These simulations aim not only to replicate experimental observations but also to extend the data. Whether spatially, temporally, or functionally, augmentation is needed for various possible uses of the components to improve the predictions of mechanical behavior. Related multimodal data are scarce, high-dimensional and a physics-based causality relation exists between observational and simulated data. In [5], a data augmentation scheme coupled with data pruning is proposed, in order to limit memory requirements for high-dimensional augmented data. This augmentation is desirable for digital twining assisted by artificial intelligence when performing nonlinear model reduction. Here, data augmentation aims at preserving similarities in terms of the validity domain of reduced digital twins. A specimen subjected to a mechanical test at high temperature is considered, where the as-manufactured geometry may impact the lifetime of the component. Hence, an instance is represented by a digital twin that includes 3D X-Ray tomography data of the specimen, the related finite element mesh, and the finite element predictions of thermo-mechanical variables at several time steps. There is, thus, for each specimen, geometrical and mechanical information. Multimodal data, which couple different representation modalities together, are hard to collect, and annotating them requires a significant effort. Thus, the analysis of multimodal data generally suffers from the problem of data scarcity. The proposed data augmentation scheme aims at training a recommending system that recognizes a category of data available in a training set that has already been fully analyzed by using high-fidelity models. Such a recommending system enables the use of a ROM-net for fast lifetime assessment via local reduced-order models.

Real-Time Data Assimilation in Welding Operations Using Thermal Imaging and Accelerated Digital Twinning Welding operations may be subjected to different types of defects when the process is not properly controlled and most defect detection is done a posteriori. The mechanical variables that are at the origin of these imperfections are often not observable in situ. In [16], an offline/online data assimilation approach is proposed, that allows for joint parameter and state estimations based on local probabilistic surrogate models and thermal imaging in real-time. Offline, the surrogate models are built from a high-fidelity thermomechanical finite element parametric study of the weld. The online estimations are obtained by conditioning the local models by the observed temperature and known operational parameters, thus fusing high-fidelity simulation data and experimental measurements.

Mechanical assessment of defects in welded joints: morphological classification and data augmentation In [15], a methodology is developed for classifying defects based on their morphology and induced mechanical response. The proposed approach is fairly general and relies on morphological operators and spherical harmonic decomposition as a way to characterize the geometry of the pores, and on the Grassman distance evaluated on FFT-based computations, for the predicted elastic response. The approach is implemented and detailed on a set of trapped gas pores observed in X-ray tomography of welded joints, that significantly alter the mechanical reliability of these materials. The space of morphological and mechanical responses is first partitioned into clusters using the "k-medoids" criterion and associated distance functions. Second, multiple-layer perceptron neuronal networks are used to associate a defect and corresponding morphological representation to its mechanical response. It is found that the method provides accurate mechanical predictions if the training data contains a sufficient number of defects representing each mechanical class. To do so, the original set of defects is supplemented by data augmentation techniques. Artificially-generated pore shapes are obtained using the spherical harmonic decomposition and a singular value decomposition performed on the pores signed distance transform.

Data Augmentation and Feature Selection for Automatic Model Recommendation in Computational Physics Classification algorithms have recently found applications in computational physics for the selection of numerical methods or models adapted to the environment and the state of the physical system. For such classification tasks, labeled training data come from numerical simulations and generally correspond to physical fields discretized on a mesh. Three challenging difficulties arise: the lack of training data, their high dimensionality, and the non-applicability of common data augmentation techniques to physics data. In [13], two algorithms are introduced to address these issues: one for dimensionality reduction via feature selection, and one for data augmentation. These algorithms are combined with a wide variety of classifiers for their evaluation. When combined with a stacking ensemble made of six multilayer perceptrons and a ridge logistic regression, they enable reaching an accuracy of 90% on the considered classification problem of nonlinear structural mechanics.

Physics-informed cluster analysis and a priori efficiency criterion for the construction of local reduced-order bases Nonlinear model order reduction has opened the door to parameter optimization and uncertainty quantification in complex physics problems governed by nonlinear equations. In particular, the computational cost of solving these equations can be reduced by means of local reduced-order bases. In [11], the benefits of a physics-informed cluster analysis for the construction of cluster-specific reduced-order bases are examined. The choice of the dissimilarity measure for clustering is fundamental and highly affects the performances of the local reduced-order bases. It is shown that clustering with an angle-based dissimilarity on simulation data efficiently decreases the intra-cluster Kolmogorov N-width. Additionally, an a priori efficiency criterion is introduced to assess the relevance of ROM-nets. This criterion also provides engineers with a very practical method for ROM-nets' hyperparameters calibration under constrained computational costs for the training phase. On five different physics problems, the physics-informed clustering strategy significantly outperforms classic strategies for the construction of local reduced-order bases in terms of projection errors.

6.4   Extension: Manifold Learning of Physics Problems Assisted by Black-Box Regressors

A priori compression of convolutional neural networks for wave simulators Convolutional Neural Networks (CNNs) are seeing widespread use in a variety of fields, including image classification, facial and object recognition, medical imaging analysis, and many more. In addition, there are applications such as physics-informed simulators in which accurate forecasts in real time with a minimal lag are required. The current neural network designs include millions of parameters, which makes it difficult to install such complex models on devices that have limited memory. Compression techniques might be able to resolve these issues by decreasing the size of CNN models that are created by reducing the number of parameters that contribute to the complexity of the models. In [7], a compressed tensor format of convolutional layer is proposed a priori, before the training of the neural network, for finite element (FE) predictions of physical data. 3-way kernels or 2-way kernels in convolutional layers are replaced by one-way fiters. The overfitting phenomena will be reduced also. The time needed to make predictions or time required for training using the original CNN model would be cut significantly if there were fewer parameters to deal with. The priori compressed models is validated on physical data from a FE model solving a 2D wave equation. Tn the considered application, the proposed convolutional compression technique achieves equivalent performance in the prediction error as classical convolutional layers with fewer trainable parameters (around 20%) and lower memory footprint.

Accelerated uncertainty quantification in impact simulations using generative adversarial networks and submodeling The analysis of parametric and non-parametric uncertainties of very large dynamical systems requires the construction of a stochastic model of said system. Linear approaches relying on random matrix theory and principal component analysis can be used when systems undergo low-frequency vibrations. In the case of fast dynamics and wave propagation, a random generator of boundary conditions for fast submodels by using machine learning is investigated in [6]. It is illustrated that the use of non-linear techniques in machine learning and data-driven methods is highly relevant. Physics-Informed Neural Networks (PINNs) are a possible choice for a data-driven method to replace linear modal analysis. An architecture that supports a random component is necessary for the construction of the stochastic model of the physical system for non-parametric uncertainties, since the goal is to learn the underlying probabilistic distribution of uncertainty in the data. Generative Adversarial Networks (GANs) are suited for such applications, where the Wasserstein-GAN with gradient penalty variant offers improved convergence results for the considered problem. The objective of this approach is to train a GAN on data from a finite element method code so as to extract stochastic boundary conditions for faster finite element predictions on a submodel. The submodel and the training data have both the same geometrical support. It is a zone of interest for uncertainty quantification and relevant to engineering purposes. In the exploitation phase, the framework can be viewed as a randomized and parametrized simulation generator on the submodel, which can be used as a Monte Carlo estimator.

MMGP: a Mesh Morphing Gaussian Process-based machine learning method for regression of physical problems under non-parameterized geometrical variability When learning simulations for modeling physical phenomena in industrial designs, geometrical variabilities are of prime interest. While classical regression techniques prove effective for parameterized geometries, practical scenarios often involve the absence of shape parametrization during the inference stage, providing only mesh discretizations as available data. Learning simulations from such mesh-based representations poses significant challenges, with recent advances relying heavily on deep graph neural networks to overcome the limitations of conventional machine learning approaches. Despite their promising results, graph neural networks exhibit certain drawbacks, including their dependency on extensive datasets and limitations in providing built-in predictive uncertainties or handling large meshes. In [10], a machine learning method that do not rely on graph neural networks is proposed. Complex geometrical shapes and variations with fixed topology are dealt with using well-known mesh morphing onto a common support, combined with classical dimensionality reduction techniques and Gaussian processes. The proposed methodology can easily deal with large meshes without the need for explicit shape parameterization and provides crucial predictive uncertainties, which are essential for informed decision-making. In the considered numerical experiments, the proposed method is competitive with respect to existing graph neural networks, regarding training efficiency and accuracy of the predictions.

References

1. N. Akkari, F. Casenave, T. Daniel, D. Ryckelynck, Data-targeted prior distribution for variational autoencoder. Fluids **6**(10), 343 (2021)
2. N. Akkari, F. Casenave, E. Hachem, D. Ryckelynck, A bayesian nonlinear reduced order modeling using variational autoencoders. Fluids **7**(10) (2022)
3. N. Akkari, F. Casenave, V. Moureau, Time stable reduced order modeling by an enhanced reduced order basis of the turbulent and incompressible 3d navier–stokes equations. Math. Comput. Appl. **24**(2) (2019)
4. N. Akkari, F. Casenave, D. Ryckelynck, C. Rey, An updated gappy-pod to capture non-parameterized geometrical variation in fluid dynamics problems. Adv. Model. Simul. Eng. Sci. **9**(1), 1–34 (2022)
5. A. Aublet, F. N'Guyen, H. Proudhon, D. Ryckelynck, Multimodal data augmentation for digital twining assisted by artificial intelligence in mechanics of materials. Front. Mater. **9** (2022)
6. H. Boukraichi, N. Akkari, F. Casenave, D. Ryckelynck, Uncertainty quantification in a mechanical submodel driven by a wasserstein-gan. IFAC-PapersOnLine **55**(20), 469–474 (2022). 10th Vienna International Conference on Mathematical Modelling MATHMOD 2022
7. H. Boukraichi, N. Akkari, F. Casenave, D. Ryckelynck, A priori compression of convolutional neural networks for wave simulators. Eng. Appl. Artif. Intell. **126**, 106973 (2023)
8. F. Casenave, N. Akkari, F. Bordeu, C. Rey, D. Ryckelynck, A nonintrusive distributed reduced-order modeling framework for nonlinear structural mechanics-application to elastoviscoplastic computations. Int. J. Numer. Methods Eng. **121**(1), 32–53 (2020)
9. F. Casenave, A. Gariah, C. Rey, F. Feyel, A nonintrusive reduced order model for nonlinear transient thermal problems with nonparametrized variability. Adv. Model. Simul. Eng. Sci. **7**(1), 1–19 (2020)
10. F. Casenave, B. Staber, X. Roynard, MMGP: a Mesh Morphing Gaussian Process-based machine learning method for regression of physical problems under non-parameterized geometrical variability (2023)
11. T. Daniel, F. Casenave, N. Akkari, A. Ketata, D. Ryckelynck, Physics-informed cluster analysis and a priori efficiency criterion for the construction of local reduced-order bases. J. Comput. Phys. **458**, 111120 (2022)
12. T. Daniel, F. Casenave, N. Akkari, D. Ryckelynck, Model order reduction assisted by deep neural networks (ROM-net). Adv. Model. Simul. Eng. Sci. **7**(16) (2020)
13. T. Daniel, F. Casenave, N. Akkari, D. Ryckelynck, Data augmentation and feature selection for automatic model recommendation in computational physics. Math. Comput. Appl. **26**(1) (2021)
14. H. Launay, D. Ryckelynck, L. Lacourt, J. Besson, A. Mondon, F. Willot, Deep multimodal autoencoder for crack criticality assessment. Int. J. Numer. Methods Eng. **123**(6), 1456–1480 (2022)
15. H. Launay, F. Willot, D. Ryckelynck, J. Besson, Mechanical assessment of defects in welded joints: morphological classification and data augmentation. J. Math. Ind. **11**(8), 18 (2021)
16. P. Pereira Álvarez, P. Kerfriden, D. Ryckelynck, V. Robin, Real-time data assimilation in welding operations using thermal imaging and accelerated high-fidelity digital twinning. Mathematics **9**(18) (2021)

17. D. Ryckelynck, A priori hyperreduction method: an adaptive approach. J. Comput. Phys. **1**(202), 346–366 (2005)
18. D. Ryckelynck, T. Goessel, F. Nguyen, Mechanical dissimilarity of defects in welded joints via Grassmann manifold and machine learning. Comptes Rendus. Mécanique **348**(10–11), 911–935 (2020)

Open Access This chapter is licensed under the terms of the Creative Commons Attribution 4.0 International License (http://creativecommons.org/licenses/by/4.0/), which permits use, sharing, adaptation, distribution and reproduction in any medium or format, as long as you give appropriate credit to the original author(s) and the source, provide a link to the Creative Commons license and indicate if changes were made.

The images or other third party material in this chapter are included in the chapter's Creative Commons license, unless indicated otherwise in a credit line to the material. If material is not included in the chapter's Creative Commons license and your intended use is not permitted by statutory regulation or exceeds the permitted use, you will need to obtain permission directly from the copyright holder.

Printed in the United States
by Baker & Taylor Publisher Services